Progressive Planet

Modern Techniques for Recycling and Reclaiming Earth

by
Alexander Hodzic

Dear Esteemed Reader,

Thank you immensely for choosing this book to join your collection. We imagine that you've already embarked on an exploration of ideas within these pages, and we couldn't be happier about it!

Now, if you find yourself chuckling, pondering, or even debating with the words in front of you, we'd absolutely love to hear about it. If you can spare a few moments to pen down your thoughts in a review, we would be as delighted as a dictionary on a spelling bee!

An Amazon review would be excellent - but hey, we're far from picky. Whether it's a scribble on the back of a grocery list, a tweet, or even a message in a bottle (though that might take a while to reach us), your feedback is gold.

Writing a review might not be as fun as a spontaneous dance-off, but we promise it'll bring grins to our faces, warmth to our hearts, and incredibly valuable insights to future readers.

With Gratitude,

Bo Bennett,
PhDPublisher
Archieboy Holdings, LLC.

Table of Contents

Introduction

Climate change is an undeniable, looming reality that we confront daily. With alterations palpable in our surroundings, there's an increased urgency to understand the adverse effects of climate change and take decisive action. Sharing that urgency, this book delves into the pivotal relationship between recycling and climate change, introducing modern, innovative recycling methods to actively participate in the ongoing fight against climate change.

The current state of our planet signals an imminent call to action. While we are becoming increasingly aware of this crisis, simply acknowledging it isn't sufficient. We are urged to transition from a society that continually consumes and disposes to one that reduces, reuses, and recycles. Throughout this book, we strive to both heighten your awareness of the climate change crisis and educate you on practical methods of recycling, emphasizing the gravity and immediacy of our situation.

Armed with the knowledge contained in the subsequent chapters, you'll learn how to reduce waste production by redirecting your consumption habits, adopting modern recycling techniques, and fully understanding the potential to find value in what is conventionally thought of as waste. As we explore the dynamic field of recycling, one that is constantly evolving and innovating, we invite you into a deeper appreciation of its intrinsic role in mitigating climate change and our responsibility toward ensuring the planet's longevity. Despite the gravity of these global issues, remember that every individual action matters. A more sustainable future is possible, let's learn how to build it together.

Chapter 1:
The State of Our Planet

As we stand on the precipice of a new era, our home - the delicate blue marble we call Earth - faces a severe ecological crisis. From the melting polar ice caps to the increasing ferocity of wildfires, the symptoms of climate change are becoming more prevalent and urgent in our daily lives. Year after year, these climatic phenomenons are breaking records and revealing a troubling trend of escalation. When we take a closer look at the roots of this crisis, we find it deeply intertwined with the relentless loss of biodiversity. It's alarming to note that species are disappearing at a rate up to dozens of times faster than the average over the past millions of years. This means extinctions are now occurring at an unprecedent rate in human history. The fragile ecosystems that uphold our survival and prosperity are weakening, impacting everything from food production to the availability of fresh water. As we delve deeper into the complexities of the issues at hand, we'll embark on an exploration of our catastrophic waste problem and how its reduction is crucial in the battle against climate change. An intimate understanding of these ailments is the first step in the right direction, as not only does it enable solution-based thinking, but also empowers each one of us individually and collectively to act towards a sustainable future.

Symptoms of Climate Change

The initial signs of climate change are increasingly perceptible and are escalating. It's akin to the fever that warns us of an underlying ailment, except it's our planet that's enduring the soaring temperatures. As the Earth's average temperature climbs, multiple disruptions and shifts ensue in weather patterns and natural processes.

Let's start with the most palpable and measurable symptom: global warming. Climate scientists around the world have noted a steady increase in the Earth's average temperature, with 19 of the 20 warmest years on record falling within the last two decades. This isn't simply about needing to wear fewer layers in winter or enduring hotter summer days - it's about the profound impact of these rising temperatures on ecosystems, agriculture, and human life.

Rising sea levels constitute another concerning symptom. As our globe heats up, two significant phenomena occur. Firstly, the polar ice caps and glaciers have begun melting at an accelerated rate, adding more water to our oceans. Secondly, warmer water expands. This thermal expansion coupled with the addition of the melted ice results in the sea levels rising around the world. Many coastal areas and islands are already grappling with the resultant flooding and erosion.

Concurrently, the world is witnessing an increase in volatile and dangerous weather events. Destructive hurricanes, typhoons, tornados, and extreme rainfall can be directly linked to climate change. These events not only pose a significant risk to human life but also have devastating impacts on infrastructure and economies globally.

Our oceans are also bearing the brunt of climate change. Increased CO_2 emissions have led to ocean acidification, a shift in pH that disrupts marine life, threatens biodiversity and impacts food chains. Warm ocean temperatures also lead to widespread coral bleaching, jeopardizing critical habitats for a plethora of marine species.

Another troubling symptom is changes in animal migration patterns. As regions become too hot or seasons shift, wildlife is forced to migrate to survive. This movement has severe ramifications on ecosystems, as it can lead to overpopulation in some areas and a decline in species in others. The extinction rate is accelerating at an alarming pace, and many species face an uncertain future.

Next, we will discuss changing vegetation zones. Warming temperatures mean vegetation is changing at both extremely high altitudes and latitudes. This change is observable in the 'greening' of the Arctic tundra as shrubs and trees encroach on traditional cold-weather plants, shifting the local ecosystem profoundly.

Besides, lands once fertile and productive are now turning into arid, dry regions due to droughts and rainfall pattern changes. This desertification not only reduces the land available for cultivation but escalates the underlying issue of food scarcity for our ever-growing population.

We must also mention wildfires, which have become rampant, ravaging landscapes from Australia to the American west. While wildfires aren't new, it's the intensity, frequency, and timing of these fires that are. It's a vicious cycle too - more fires mean more carbon released into the atmosphere, which in turn exacerbates climate change.

The rate of permafrost thaw is another symptom worth discussing. As the Earth's temperature increases, the traditionally frozen ground in the Arctic begins to thaw, releasing massive amounts of methane, a potent greenhouse gas that accelerates global warming.

Then there's the reduced snowfall and shrinking snowcaps, particularly in mountain regions. Changes in snowpack levels impact the timing and quantity of water flow to communities, agriculture, and ecosystems downstream. It's a reverberating impact, with consequences for water availability, food production, and even energy generation.

A phenomenon worth noting too is the alteration in disease patterns. Warmer temperatures and changes in rainfall patterns have led to the spread of vector-borne diseases such as malaria, dengue, and Lyme disease to areas that were previously unaffected.

Indeed, not forgetting to mention the increasing acidity of our soils. Elevated CO2 levels lead to a reduction in the nutritional quality of crop plants, simultaneously increasing the growth of weeds and pests. This change threatens global food security significantly.

The final, yet significant, symptom is the change in human behavior and activity. Rising temperatures can affect people's physical health and cognitive function, lead to displacement from extreme weather events or sea level rise, and even cause conflicts over increasingly scarce resources.

These symptoms of climate change serve as an urgent call to action for humankind. As we observe the Earth's shifts and disruptions, it's crucial to understand the role we play in preserving our planet for future generations.

Biodiversity Loss

The world as we know it is in a constant state of evolution, but human activities have been accelerating changes in a way that's proving harmful to many species. This rapid and multifaceted change in biodiversity, unfortunately, isn't receiving the attention it deserves. This section is dedicated to shedding light on the scourge of biodiversity loss, a critical issue unfolding in the backdrop of climate change and increased waste generation.

Let's start by understanding biodiversity. Biodiversity is the variety of life on Earth, in all its forms and all its interactions. It is the fabric of life, from the smallest microorganisms to the largest mammals; it encompasses the diversity within species, between species, and of ecosystems. The multitude of life forms on our planet not only adds color and richness to our lives but also underpins the healthy functioning of the world.

However, Planet Earth is now experiencing a swift and disturbing loss of biodiversity. Scientists have noted that extinctions are occurring at a rate much higher than the estimated background extinction rate, with numerous species disappearing before we even discover them. This higher rate of extinction is often linked to human activities, particularly those related to climate change, deforestation, pollution, and unsustainable consumption.

The loss of biodiversity is a pressing issue for several reasons. Firstly, every species plays a specific role in our ecosystem. The disappearance of any species disrupts the balance, with subsequent changes rippling out and affecting various other species and ecosystem processes. Be it plant species that hold the soil together, reducing the rate of erosion, or a predator that controls the population of a prey species, the role of each organism is irreplaceable and essential to the functionality of the ecosystem.

Secondly, the loss of biodiversity threatens our food security. We rely on a variety of species for our food, and the loss of a single species can affect our agricultural systems. Be it the numerous plant species grown as crops or the bees that pollinate these plants, maintaining biodiversity is crucial.

Biodiversity also holds the key to advancements in various fields, including medicine. Many major breakthroughs, including cancer treatments and pain relief medications, have been derived from plants and animals. The untapped potential within the diverse range of species on Earth is enormous, and the loss of these species is the loss of potential cures and advancements.

Additionally, biodiversity loss has economic implications. Consider our reliance on ecosystem services like clean water, breathable air, and climate regulation—all these are directly dependent on healthy and diverse ecosystems that we are rapidly losing.

Climate change and biodiversity loss are intrinsically linked. Rising temperatures, changing precipitation patterns, and intensifying weather events all impact ecosystems, upsetting the delicate equilibrium that sustains different species. In turn, biodiversity loss exacerbates climate change because diverse ecosystems capture and store more carbon dioxide than do degraded ones.

Moreover, human activity and unsustainable practices have greatly intensified biodiversity loss. Habitat destruction, typically resulting from deforestation for agriculture, urban development, or logging, is the leading cause of biodiversity loss. Other significant contributors include overexploitation of resources, pollution, invasive species, and climate change.

The loss of biodiversity isn't just an environmental issue; it's a developmental, economic, security, social, and moral issue as well. To halt the loss of biodiversity, urgent and unprecedented changes are required in how we live, work and consume.

As we discuss ways to combat climate change and approach waste management throughout the book, consider how these practices can contribute to preserving biodiversity. For instance, lessening our reliance on single-use plastics can reduce pollution in our oceans and save countless aquatic species. Similarly, moving toward a circular economy helps to preserve resources, reduce waste, and lessen the strain on our natural environment.

Mitigating biodiversity loss isn't just a task for environmental scientists or policy-makers; it is an undertaking in which we all must participate. In doing so, not only will we be preserving the vibrant array of life on Earth, but we'd also be safeguarding our futures and those of coming generations. Now is the time to act. Every day counts and each individual effort matters.

In conclusion, biodiversity loss is a grave issue that has been overlooked for too long. While unplanned urbanization, deforestation, pollution, climate change, and unsustainable consumption all contribute to the issue, we now have the opportunity, and indeed the responsibility, to halt and reverse the loss of biodiversity through informed actions and conscientious living.

Chapter 2:
The Waste Predicament: Identifying and Reducing our Waste

Having explored climate change and its impact on biodiversity in the preceding chapter, we now find ourselves faced with a momentous challenge: addressing the burgeoning problem of waste. Identifying and reducing waste is not just a matter of keeping our streets clean. It's about preserving the prospects of our ecosystems and securing our future against the dire consequences of unchecked consumption and waste generation. The complexity of this issue stems from the multiple types of waste we produce daily, each trapping us in an ecological quagmire. Predominantly, food and plastic waste take the lead, quietly contributing to climatic drifts and ecosystem instability in ways that we often overlook. The notoriety of food waste is largely on account of its invisibility; many don't consider the sheer resources involved in producing food, and only perceive the tangible, uneaten produce as 'waste'. Plastic, on the other hand, is a highly visible villain. Its resilience to degradation and sheer abundance have set our environment on a perilous path. Our immediate response to these two waste streams, both with their unique challenges, will inescapably shape our global waste outlook in the forthcoming years. In our journey to understand and tackle waste, we will delve into the many avenues we have to truncate this growing problem, including new economic models, updated waste technology, and incentivizing personal responsibility.

Food Waste: The Invisible Culprit

Continuing from our exploration of various types of waste, let's now dive into the riveting world of food waste; an often overlooked aspect in the vast landscape of waste management. Food waste is an invisible culprit with significant environmental impacts. While we, as a society, have given extensive attention to plastic and other forms of physical waste, the issue of food waste has silently grown to an alarming scale and demands urgent twofold attention: reducing the waste and recycling it effectively.

Allow us to paint a picture. The United Nations estimates that one-third of global food production, around 1.3 billion tons per year, is wasted. This is equivalent to more than $1 trillion worth of food, thrown away or lost in various stages from farm to plate. It's staggering, isn't it? However, the economic loss is just one facet of the issue. This degree of food wastage also equates, regrettably, to a colossal squandering of environmental resources used in food production, including land, water, energy, and labor. Together, it contributes around 8% of the total global greenhouse gas emissions.

The issue of food waste extends its tentacles beyond mere numbers. It mirrors our systemic inefficiencies, consumeristic tendencies, and a gross lack of respect for resources. When food is discarded, it's not just the food itself going to waste, instead, we're squandering an entire chain of resources from farm to fork. For example, to grow the food, we need water, fertile land, and labor. Food processing requires even more resources. This chain of waste gets worse as we consider what happens to the wasted food in landfills.

Once food waste reaches a landfill, it's left to break down anaerobically, a process that releases methane—a potent greenhouse gas, with a global warming potential many times greater than its more infamous cousin, carbon dioxide. These methane emissions from rotting food waste in landfills are a significant contributor to the climate crisis. It's an ironic cycle, where we are effectively creating food to heat the planet.

So, how did we get here? Food waste is largely systemic and can manifest at various points in the supply chain. Overproduction at the farming stage, improper storage, damage during transport, sales promotions encouraging over-purchasing, prepared food not being consumed, and consumer habits of wasting leftover food all contribute to the food waste crisis.

Furthermore, there is a lack of public awareness about the scale of this problem and insufficient nationwide infrastructures and legislation in place to manage food waste efficiently. A big chunk of food waste could be avoided if everyone understood the scale and consequences of the problem, and simultaneously, governments initiated adequate systems and measures to manage it.

At this juncture, it's also crucial to understand the difference between avoidable and unavoidable food waste. Avoidable food waste is the edible food that could have been consumed but ended up in the waste bin for various reasons. It includes whole foods and edible parts of foods that were discarded such as crusts of bread, and leftover meals. On the other hand, unavoidable food waste includes inedible parts like coffee grounds, eggshells, fruit peels, and bones.

While efforts should be made to minimize avoidable food waste, strategies need to be in place to manage unavoidable waste such as composting, energy recovery, and valorization—ascribing value to by-products. In fact, significant resources can be extracted from unavoidable food waste, closing the loop in a circular economy style that recycles nutrients and reduces environmental impacts.

We now have innovative, modern ways to reduce and recycle food waste. For example, anaerobic digestion is a process where microorganisms break down organic materials, such as food waste, in an oxygen-free environment, producing biogas. This biogas can be used to generate heat and electricity or can be upgraded into biomethane—a renewable substitute for natural gas.

Another example of cutting-edge technology is Black Soldier Fly Larvae (BSFL) composting: the larvae of this insect species can quickly breakdown organic waste, including food scraps, into rich compost. The larvae themselves can also be harvested as a protein source for livestock feed—a remarkable, reductive, and regenerative cycle.

In reviewing these technologies, it's clear there are meaningful opportunities for change. But to really curb food waste and its environmental impacts, we need a multi-pronged approach that combines policy interventions, technological advancements, and shifts in societal norms and behaviors. That begins with everyone understanding and acknowledging this invisible culprit.

We, as individuals, also have a compelling role to play. We need to be conscientious about our shopping habits, plan our meals, properly store our food, and creatively use leftovers. By doing this, we not only reduce waste but also save money. It's a win-win.

The issue of food waste is enormous, its consequences dire. We should treat food not only as a necessity but also as a valuable resource that carries the weight of environmental impacts, labor, and love. Food waste—the invisible culprit, is our collective burden, and tackling it will require reinvention at every level, from individual households to large-scale industries and national policies.

Plastic Pollution and Waste Overload

The rampant use and discard of plastic has led to a unique issue that our planet faces today - plastic pollution and waste overload. From industries to households, the prevalence of plastic is palpable and pressing. A material so durable that it resists decomposition, yet so disposable that it has become synonymous with waste.

The scenario is grim. Plastics comprise up to 90% of all waste floating on the ocean's surface, with about 46,000 pieces of plastic per square mile. It's estimated that the average American throws away approximately 185 pounds of plastic per year. The plastic bags you see fluttering about in trees, stream beds, seas, and oceans will endure for hundreds of years.

The disintegration of plastic is not a biodegradable process but rather a fragmentation. What this means is that these plastic items don't biodegrade and return to nature. Instead, they gradually just become smaller, tiny toxic bits of original material, referred to as microplastics. It's easy to perceive the problem posed by a discarded water bottle or single-use plastic bag, but microplastics pose an 'invisible' threat.

Microplastics are insidious. They permeate every level of ecosystems, harming wildlife and eventually ending up in our own bodies. Microplastics are found everywhere – from the deepest ocean trenches to the remotest Arctic ice. They are ingested by wildlife and transported up the food chain, ultimately, into the human diet.

Then there's the issue of waste overload. Our current handling of plastic waste isn't sustainable. Approximately eight million metric tons of plastic are thrown into the ocean each year. In the United States alone, only 9% of plastic is recycled, whereas 91% of plastic isn't recycled. Our landfills are brimming, and our recycling systems are under immense strain.

There's another alarming side to this—greenhouse gas emissions. Plastic production utilizes around eight percent of the world's oil production, emitting considerable amounts of greenhouse gases. Moreover, when plastic is improperly disposed, it often ends up being burned in open air conditions, further leading to greenhouse gas emissions and harming human health.

The graphic image of a sea turtle ensnared in a discarded fishing net or a seagull strangled by a six-pack ring are sobering examples of the toll plastic pollution is taking on wildlife. Over 1 million marine animals— including mammals, fish, sharks, turtles, and birds— are killed each year due to plastic debris in the ocean.

Societal convenience has come at an incredible ecological cost. Plastics, particularly single-use plastics that we use daily such as grocery bags, straws, and water bottles have inundated our earth and oceans. The need to transition away from single-use plastics to more sustainable options is increasingly apparent.

Pointing fingers at industry conglomerates will change little unless we couple it with tangible individual changes. After all, the plastic problem is everybody's problem. Economic, environmental, and societal implications exist alike. Better waste management and recycling, reduction in plastic use, and investment in newer, less harmful materials are the solutions that we can work towards.

While the challenge of plastic pollution and waste overload is daunting, it is not insurmountable. By making conscious decisions about the products we purchase, we lay the groundwork for creating a less plastic-dependent society. As we reduce, reuse, and recycle, the fallout from the plastic problem can be mitigated.

In our homes, we can implement small yet significant changes. Opting for reusable shopping bags, ditching the plastic wrap for beeswax covers, shopping in bulk to reduce packaging, refusing single-use items like straws and cutlery are measures that collectively have an impact.

We also need to advocate for large scale solutions, including laws to limit or ban certain types of plastic, like lightweight single-use shopping bags, straws, and Styrofoam cups. Companies must be encouraged, or even legislated, to take responsibility for the lifecycle of their products, including the post-consumer phase.

No step is too small. Every action holds potential. Progress may be slow, but the shifts are tangible. Together, we can navigate out of this storm of synthetic waste. Because we can't let the tale of plastic become a tragedy of our planet.

As we move forward in this discussion on our waste predicament, let us keep in mind our shared responsibility. In the following sections, we'll delve deeper into consumption patterns, the economic systems underpinning them, and the role of recycling within these systems. We'll explore modern methods of recycling, look at how we can make better personal choices, and examine ways of breathing new life into used materials. The path maketh the traveler, so let us tread lightly, wisely, and with vigilance.

Chapter 3: Reassessing Our Relationships with Consumption

This part of our exploration necessitates a deep dive into the double-edged phenomenon of consumption, tackling both the linear and the circular economic models. The essence of the linear economy lies in the 'take-make-waste' methodology which, though market-friendly, adds momentously to the climate plight. This 'one-use' model is deeply entrenched in our current economic framework and emphasizes consuming without considering the waste generated. The direct contrast is the circular economy, where resources are maintained, recycled, and reused with minimal environmental impact. Why is this model vital for our survival? It fundamentally alters the way we deal with waste, positioning recycling at the center-stage. All products, post-initial use, are redefined as 'new resources', fostering constant resource circulation. Moreover, beyond the economic lens, our actions must reflect our awareness of climate change implications. Both of these frameworks distinctly sculpt our approach to consumption. By understanding them, we ought to embed conscientious practices into our daily routines, moving towards sustainable consumption and climate positivity.

The Linear Economy

Modern consumption patterns primarily exist within a paradigm known as the linear economy. This structure predominantly views the economic cycle as a straight line, or to put it in simpler terms, it espouses a 'take-make-dispose' philosophy. This kind of strategy appears sound in the short term, yet over time it has significant ramifications on the natural environment.

Our economic model currently relies heavily on extensive resource extraction. As industries continue to use raw materials to meet demand, these resources, whether mineral, fossil, or biological, slowly diminish. Industrial production is heightened to match consumer demand, leading to an increase in greenhouse gas emissions, as well as other forms of pollution.

In this structure, products tend to display a short lifespan. They are generally designed to last until the next upgrade or model is released. Once obsolete, these goods typically head to landfills or incinerators. The disposable nature of products has significant environmental implications by creating overwhelming amounts of waste while also further depleting natural resources.

The by-products of this economic model affect biodiversity as well. The immense quantities of waste produced often disrupt natural ecosystems, sometimes irreparably. Depletion of resources also leads to the destruction of habitats, putting many species at risk.

One of the stark features of a linear economy is the maintenance of high levels of demand. Advertisements and consumer culture constantly encourage the acquisition of new products, promising higher status, greater convenience, or simple novelty. This cycle of constant production and consumption exacerbates the existing strains on the planet's resources.

Energy consumption in a linear economy is another talking point. To produce goods and services, we rely primarily on finite sources of energy such as coal, oil, and gas. Overuse of these fuels contributes to air pollution and climate change, while simultaneously eroding energy security for future generations.

Another characteristic worth mentioning is the creation of jobs through consumer spending. While it may seem positive, it often traps individuals in a cycle of earning and spending, discouraging them from questioning the system's long-term impacts. It instead promotes a rampant consumerism mindset, further fueling the 'take-make-dispose' system.

The linear economy's impact on the planet is concerning. It contributes heavily to climate change, environmental degradation, as well as biodiversity loss. It also compromises the well-being of future generations as the depletion of resources and energy reserves continues unabated.

Beyond environmental impacts, the linear system raises social concerns too. The rampant consumerism culture can often lead to economic inequality. Those at the highest points of consumption are typically the wealthiest, yet the environmental costs fall on the global community, including those least responsible for the problem.

The linear economy, while seemingly functional, has proven to be fundamentally unsustainable in the context of our planet's finite resources. Ultimately, it fails to respect the need for balance between human activities and the natural world. This imbalance inevitably leads to a gradual depletion of Earth's resources and harm to its ecosystems.

Moreover, the linear model significantly contradicts the established laws of nature. In our natural world, nothing is considered waste. Every organism plays a part in a constant cyclical process of growth, death, and decay, contributing to the nourishment of new life. This contrast blatantly reveals the inadequacy of a linear economy in the long term.

Perhaps now is the time to question the wisdom of sticking to this status quo. The urgency of the climate crisis, coupled with the ethical implications of resource depletion and environmental degradation, necessitates a change in our consumption patterns. But is there an alternative to this harmful economic model? As it turns out, there is: the idea of a circular economy.

The rhythms of nature serve as an ideal blueprint for this sustainable and productive economic model. In the chapters to follow, we'll delve deeper into the circular economy, understanding its importance, and examining the role recycling plays in promoting a more sustainable future. We'll learn how we can break the linear cycle, make the most out of our resources, and work towards a healthier planet for all.

Despite the dominance of the linear economy, we have seen encouraging signs of change. Innovative policies and practices are gradually shifting the focus from endless consumption to sustainable, circular principles. This movement towards a more sustainable model requires collective effort — from individuals, businesses, governments, and everyone in between. Are we ready to embark on this journey towards a healthier planet? Let's dive in.

The Circular Economy

As we delve further into the heart of our relationship with consumption, a particular model known as the 'circular economy' emerges as an increasingly relevant concept. A distinct hail from the traditional linear economy discussed in the previous section, the circular economy aims to reshape our understanding of growth and development.

The principal idea behind the circular economy is to go forward by looking backward. It strikes a chord with the natural cycle of life, where everything has a life cycle, and yet nothing is truly disposable. There is a particular focus on "closing the loop" through designing out waste and pollution, keeping products and materials in use, and regenerating natural systems. This model signifies a shift away from our 'take-make-dispose' pattern of growth towards a circular approach.

The circular economy is not merely a blueprint for environmental sustainability but a robust strategy for economic resilience too. By decoupling economic activity from the finite supply of natural resources, it ensures a secure future underpinned by sustainable growth. But this transition from linear to circular is as equally challenging as it is transformative.

While a perfectly circular economy currently remains an idealistic ambition, multiple elements can bring us closer to this goal. One of the primary drivers is the recycling industry, which has the potential to transform our current throwaway culture over time by ensuring that resources keep flowing within the economic system.

The role of recycling in a circular economy is multifaceted; it saves raw materials by providing industry with secondary resources, reduces energy consumption and decreases environmental damage caused by extraction processes. Fewer natural resources are used, less waste ends up in landfills, and greenhouse gas emissions are broadly reduced. It's a win-win-win situation.

However, recycling in a circular economy is not limited to materials such as plastic, paper, or metal. We're talking about a larger vision where everything - products, components, and materials at every level - is recycled. It's about redirecting resources back into the economic loop, without losing their value or utility.

The idea is to go from cradle to cradle rather than from cradle to grave. We need to create a system where waste is not just minimized but eliminated. Everything is part of a continuous cycle of use and reuse. The result is a strong, healthy economy, and a healthy planet.

For a circular economy to succeed, however, every link in the value chain must participate. This begins with product design. Designers must prioritize recyclability and longevity in their creations. Manufacturers should then utilize these designs to create durable, non-toxic products that are as energy-efficient as possible. Retailers can promote these products, and consumers can choose them over less sustainable alternatives.

At the same time, policymakers can play a significant role in the transition to a circular economy. They can provide incentives for businesses that prioritize sustainability and penalize those that don't. They can invest in research to refine recycling technologies and transform waste management infrastructures. And they can use education to emphasize the importance of recycling and instilling in people the values of a circular economy.

It's also important not to overlook the power of individual actions. By making conscious choices about our consumption habits, we all can contribute to the circular economy. When we choose to buy less but better, repair instead of replace, or patronize businesses that prioritize sustainability, we are helping to close the loop.

As we move forward, the goal should not only be recycling as we know it today. The aim should be to shift the whole economic system toward one that's circular, a system that's regenerative by default. And with ongoing technological advancements, breakthroughs in material science and growing environmental consciousness, a shift to the circular economy appears not just possible, but probable.

In the end, the circular economy offers a ray of hope in our pursuit of sustainability. As an environmental strategy, it reduces waste and curbs overconsumption. As an economic strategy, it encourages innovation and improves resource efficiency. Together, these benefits contribute to a more sustainable, resilient, and regenerative economy – an economy capable of counterbalancing the strain we've put on our planet.

We must remember that our survival and prosperity as a species are intimately tied to the health of our planet. It's time we align our economic ambitions with the fundamental principles of life on earth. That's where the circular economy, with its transformative potential, comes into play. As we continue this journey, let's hope the circular economy isn't merely a trend but becomes the new norm, a staple in our shared quest for a sustainable future.

Importance of Circular Economy As we continue our exploration of climate change and its relationship with waste management, we arrive at the heart of the matter - the circular economy. The circular economy highlights a groundbreaking approach to consumption and waste, wherein resources are kept in use for as long as possible, and waste becomes a thing of the past. Put simply, in a circular economy, we make, we use, and then we remake, creating a continuous loop of sustainable productivity.

This shift towards a circular economy is of immense importance, not just as a mechanism to reduce waste, but as a tool to fight climate change as well. Linear economies, with their focus on producing, consuming and disposing, simply cannot be sustained in the long term without catastrophic environmental consequences. The incessant extraction of finite resources contributes to biodiversity loss and the aggravation of climate change, among other negative impacts.

In contrast, a circular economy addresses these issues head-on. By keeping resources in use longer, we decrease the need to extract more, reducing environmental strain. By designing out waste and pollution, we minimize the damage done to our planet. And by regenerating natural systems rather than undermining them, we build towards a more sustainable future. The circular economy's potential to lower greenhouse gas emissions and reduce reliance on resource extraction underscores its importance in the fight against climate change. It is, in essence, a significant solution that facilitates environmental restoration and prevents further degradation of our planet.

Role of Recycling in the Circular Economy serves as a crucial pivot point in the transition from a linear economy to a more sustainable, regenerative circular economic model. In a circular economy, resources are kept in use for an extended period through a closed-loop system that prioritizes reusing, repairing, refurbishing, and recycling products and materials. This provides a stark contrast to the traditional linear system, where we extract resources, make products, use them, and then discard them, resulting in wastage and environmental pollution.

Specifically, recycling plays a key role in this circular framework by turning waste into valuable resources that can be reintegrated into the economy. The goal is to convert waste materials into new materials or objects, preventing unnecessary extraction of fresh resources, reducing energy usage, reducing air and water pollution, lowering greenhouse gas emissions and conservely contributing significantly to climate change mitigation. This approach not only promotes the idea that waste is a misplaced resource that can be reclaimed, but also underscores the very essence of recycling; that it is an integral part of an economy where nothing should be wasted.

Moreover, recycling in the circular economy cascades a myriad of economic benefits. It fuels economic development by creating job opportunities, fostering innovation, and driving competitive advantage. By extending the product lifecycle and improving resource efficiency, recycling also aids in reducing costs associated with waste management and helps in managing resource scarcity. Therefore, embracing recycling in the circular economy can not only aid in creating a more sustainable and resilient planet but also stimulate economic growth, making it a win-win solution for environmental and economic challenges.

Chapter 4:
Understanding Recycling

The initial step towards fully grasping the essence of recycling begins with understanding the traditional approach. The tried-and-true methods, governed by the mantra - "reduce, reuse, recycle," paint a simplistic process, but tend to overlook the complexities inherent in waste management. Conventional recycling focuses primarily on converting materials like paper, glass, plastic, and metals into completely new products. This method, while commendable, does have its set of limitations. The levels of quality reduction of recycled materials, energy consumption during the process, and hindrance posed by contamination all significantly impact the overall efficacy of traditional recycling.

With the snowballing volume of waste being produced, there is an evident need to revolutionize the recycling process. Today, we're witnessing a surge in modern innovations in recycling. New, groundbreaking technologies are steadily surfacing to counteract the limitations of the conventional approach. These technologies showcase an extensive range of capabilities, from distinguishing non-recyclable items to recycling products at a molecular level. With an ever-increasing focus on sustainability, these modern innovations promise to reshape recycling for future generations. In the following chapter, we'll delve into the specifics of these novel techniques and the endless possibilities they herald for an eco-friendly future.

The Traditional Approach to Recycling

Historically, recycling has acted as a significant buffer between waste and landfills, providing a means of lessening the overall environmental impact caused by discarded materials. The traditional approach to recycling has centered upon a set cycle - collect, sort, process, manufacture, and sell.

The initial step involves collecting recyclables. Local governments or waste management corporations curate this process, coordinating curbsides, drop-off schemes, or 'buy-back' sites. The objective is to accumulate a sufficient amount of waste for the subsequent recycling steps.

Post-collection, the gathered materials are then transported to a materials recovery facility (MRF) for sorting. This process is primarily assisted by people, although machines are often used for heavier objects such as glass bottles or metal cans. The sorting process is meticulous, separating materials into distinct types like plastic, paper, glass, or metal.

Each collected resource is then processed according to its nature. Metals are shredded and melted, plastics are thermally treated, and paper products are washed and bleached. The process aims to eliminate impurities and repurpose the waste into a state that mimics its original form.

The processing stage of recycling aggregates all the separated and cleaned recyclables into 'commodities' - raw materials that could be used to replace virgin resources in the manufacturing of new products. For example, aluminum cans could be melted into sheets of aluminum. It's important to note, however, that the quality of recycled materials doesn't often match the caliber of newly explored resources.

The processed material is not the final product in the recycling flowchart. Instead, industries use them to manufacture new products, ranging anywhere from paper towels to park benches. This manufacturing stage is the heart of the traditional recycling process, creating demand for recycled materials and cementing recycling as an integral part of the waste management landscape.

Finally, like any other product, these items go to market for consumers to purchase. The cycle, ideally, continues infinitely, with every phase playing its part in conserving our natural resources by reusing our manufactured goods.

One of the defining aspects of traditional recycling is its reliance on 'Single Stream Recycling'. Its simplicity lies in the fact that all recyclable material is mixed in collection and sorted later at a recovery facility. This method encourages greater participation due to the sheer convenience, boosting overall recycling rates.

However, while traditional recycling has played an unquestionable role in waste management, it's not exactly flawless. The very convenience of the single-stream method plays into its most significant drawback - contamination. When different types of recyclable materials are mixed together, they can become tainted by one another, reducing the overall quality of the recovered materials and even making them unrecyclable in some cases.

Furthermore, the traditional approach has long fought against the economics of recycling. The cost of collecting, sorting, and processing recyclables can be prohibitively expensive, especially when the price of virgin materials is comparatively cheap. This monetary disadvantage has stagnated the growth of recycling infrastructure, hindered by the lack of financial incentive.

Moreover, the scope of traditional recycling is also limited. As it stands, not all materials are recyclable. Plastics, in particular, pose a significant challenge due to the sheer variety of types and subtypes. Furthermore, different jurisdictions may have varying rules regarding what is and isn't accepted into their recycling programs.

Lastly, the emphasis on recycling might also inadvertently detract from more sustainable practices such as waste reduction and reuse. By providing an outlet for waste products, recycling may inadvertently encourage the continuation of a consumerist, "throw-away" culture.

While this classic approach has had its notable achievements, we are faced with an escalating need for more innovation in the field witnessed by the rising tides of waste and the encroaching threat of climate change. The necessity for adaptation and evolution in our recycling practices has never been more urgent.

Limitations of Traditional Recycling

Traditional recycling, intended to address waste issues, has indeed played a vital role in the management of resources. However, it is far from perfect. As beneficial as it has been, traditional recycling also has its flaws and inherent limitations requiring thoughtful reconsideration and innovative solutions.

The first limitation lies in the technical process of recycling. Traditional recycling often takes a 'downcycling' approach where materials, once recycled, suffer a significant decrease in original quality. We can't ignore the fact that every time plastic is recycled, for instance, its quality diminishes, leading to a limited recycling lifespan. Subsequently, the need for virgin materials to add to the recycled plastic to maintain its usefulness arises, undermining the goal of reducing raw material extraction.

A compounding issue is contamination. This occurs when non-recyclable materials are mixed with recyclables, or when different types of recyclables are mixed. Sadly, contamination can render entire batches of materials unrecyclable, leading to unwanted waste. Although there are rigorous sorting processes, mechanical and manual, they aren't flawless and contamination remains a stubborn obstacle.

Moreover, not all materials are recyclable. Chewed-up bubble gum, used napkins, coffee cups lined with plastic, and certain types of plastics, like polystyrene and polyvinyl chloride, are among the items that typically get thrown into recycling bins, but can't be processed properly. The inability to recycle these materials underscores a significant limitation in the current recycling infrastructure.

Another critical limitation of traditional recycling is its energy-intensive nature. The process can involve crushing, melting, and reforming materials, which require substantial amounts of energy. And while recycling does generally use less energy than creating new materials from scratch, there is still a significant environmental cost associated with the process.

We also can't ignore the economic aspect. Often, recycling is not cost-effective when compared to the production of virgin materials, largely because of low oil prices. When the price of oil drops, so does the cost of making new plastic, and suddenly it's cheaper for manufacturers to use new, rather than recycled, plastic in their products.

Moving on, there's the issue of the sheer volume of waste. Despite best efforts, the rate at which recyclable waste is being produced far outstrips the rate at which it can be processed. Without a fundamental shift in how we produce and dispose of materials, we will always be trying to catch up.

Furthermore, recycling infrastructure is unevenly distributed worldwide, with some countries having well-developed systems, while others have little to none. This gap often leads to the trans-boundary movement of waste, creating additional environmental and social issues, including carbon emissions from transporting waste and the exploitation of countries with weaker environmental regulations.

The process of recycling itself also has its shortcomings. For instance, recycling paper requires the use of bleach to remove ink, which pollutes waterways. Additionally, heating plastics during recycling can release hazardous chemicals into the air. This points to an inherent contradiction in the recycling process: it's meant to reduce environmental harm but can also contribute to it.

Attempting to scale up traditional recycling methods can lead to inadvertent increases in greenhouse gas emissions. This is due to a combination of the high-energy demand of the recycling process and the emissions produced in the collection and transportation of materials.

The disposal of materials that cannot be recycled is an additional challenge. Landfills, even those designed to be environmentally friendly, still pose threats to natural ecosystems and contribute to emissions of greenhouse gases. The search for sustainable, long-term solutions for these materials remains an ongoing struggle.

Interestingly, recycling can also lead to a false sense of environmental responsibility, in which the action of recycling allows for the justification of overconsumption. This "feel good" element of recycling doesn't encourage the reduction in our consumption habits.

In summary, while traditional recycling processes do have their advantages and made significant strides in waste management, the limitations and imperfections are substantial. As these issues suggest, while recycling is still an essential component of waste management, it can't bear the burden alone.

In addressing this, we don't need to discard traditional recycling altogether. Rather, it is about progressing beyond its limits, examining the root causes of waste, and working towards comprehensive, sustainable solutions that integrate various approaches. This mission feeds our focus on innovative modern recycling methods and the promotion of comprehensive waste reduction strategies.

Modern Innovations in Recycling

As we continue our exploration of the world of recycling, it is essential to understand that recent innovations have revolutionized this world. Technological advancements have bolstered the efficacy and efficiency of waste management, enabling us to recycle materials that were once deemed unrecyclable.

Noteworthy among these groundbreaking technologies is automated sorting. In previous decades, material recovery facilities depended heavily on human labor to sort out various recyclables from the waste stream. Today, infra-red scanners, magnetic fields, and advanced robotics have made it possible to separate waste materials accurately and effectively at high speeds, greatly improving the efficiency of recycling programs. These automated systems can identify recyclables based on their material properties, even separating different types of plastics to ensure they are correctly processed.

Another important innovation is bioleaching. This method uses microorganisms to extract precious metals like gold, silver, and copper from electronic waste, a type of waste that is not only becoming increasingly prevalent in our digital world but also highly problematic due to its complex and hazardous components. This innovative tech represents a huge leap forward in e-waste recycling, offering an environmentally friendly alternative to conventional recovery methods that typically employ toxic chemicals or need excessive amounts of energy.

Our fight against plastic waste, one of the most significant contributors to environmental degradation, has found an ally in the form of enzymes aptly dubbed "plastic-eating." These naturally occurring organisms break down different kinds of plastics, transforming them into their constituent parts that can then be reused to produce new plastic materials without the need for any fossil fuels.

On a larger scale, waste-to-energy technologies have also been attracting global attention. As the name suggests, these techniques convert different types of waste, particularly non-recyclables, into heat, electricity, or fuel through processes such as incineration, gasification, and anaerobic digestion. These techniques provide a viable recycling pathway for otherwise non-recyclable waste, offering a dual benediction of waste reduction and renewable energy production.

Surprisingly, carbon absorption technology is emerging as a novel approach to recycling. This technology is designed to capture and convert carbon dioxide, a notorious greenhouse gas, into useful products, simultaneously mitigating climate change and generating valuable commodities. This development epitomizes the kind of circular economy we are gradually propagating.

Although less technical, community-based recycling programs have also seen innovative changes that enhance local waste management. For instance, some programs have shifted into a pay-as-you-throw model that charges residents based on the amount of disposal waste they generate, encouraging them to recycle more and dispose of less.

Perhaps creativity and invention shine brightest in the arena of product design. Products are being designed to be recyclable from inception, taking into consideration the product's entire life cycle. This design paradigm, known as "design for recycling," ensures that recycling factors are accounted for during a product's design phase, simplifying eventual recycling processes and reducing waste generation.

In the construction industry, innovations include green concrete. It's made by replacing a portion of the traditional concrete component - cement - with waste by-products, reducing carbon emissions drastically without affecting the strength and function of the resulting material.

Virtual waste tracking, another significant innovation, employs the prowess of blockchain technology to track waste globally, ensuring transparency and accountability in waste management streams. This fosters trustworthy recycling processes and contributes to data availability, which ultimately facilitates better waste policy configurations by government entities.

Innovations in recycling don't just stop with technologies but also extend to business model innovations. Waste-as-a-service and product-as-a-service models are redefining waste streams by centralizing waste management processes or transforming products into services to minimize waste generation, respectively.

Lastly, we have to give credit to the proliferation of recycling apps. These digital platforms offer convenient platforms for people to monitor their waste generation, educate themselves about recycling, and connect with local recyclers. This digital empowerment of individuals serves as a testament to how technology continues to democratize recycling, making it more accessible to everyone.

In conclusion, modern innovations in recycling are invaluable, transforming the way we manage waste and throwing us a lifeline in our fight against climate change. While we've come a long way in repurposing and reusing our waste, there's still much that we can do. Each of these innovations provides us with not just a way to manage our waste but also a blueprint for the kind of society in which we want to live – one that respects resources, minimizes waste, and acts responsibly for the planet's future.

Chapter 5:
Modern Recycling Techniques

As we navigate the dense jungle of waste, prevalent innovations in recycling techniques have emerged as a beacon of hope. These advancements are particularly effective in handling materials that have plagued traditional recycling methods. Advanced plastic recycling technology is one such development, employing chemical processes to break down plastic waste into its foundational polymers. The resultant raw materials, reusable for manufacturing new plastic products, can drastically reduce our reliance on virgin plastics and assist in mitigating the environmental damage wrought by excessive plastic waste. In parallel, industrial composting techniques have redefined how we handle organic waste. By maintaining precisely regulated environmental conditions, waste decays more efficiently, promoting widespread use of organic waste as rich, nutrient-filled compost, reducing the dependency on chemically laden fertilizers. Recycling of electronic waste, or e-waste, has also seen a welcome evolution. By utilizing innovative shredding and sorting technology, valuable metals can be safely and effectively reclaimed from discarded electronics. This not only prevents hazardous waste from infiltrating our ecosystems but also addresses the rising demand for rare and precious metals. These breakthroughs in modern recycling practices are poised to have profound impact, potentially reshaping our global waste management strategies while combating the accelerating threat of climate change.

Advanced Plastic Recycling Technology

The conventional means of recycling plastics have been primarily mechanical, resulting in products of lower quality and value. With the advancement of technology in the 21st century, newer, more sophisticated methods have been discovered to overcome this challenge. One such method is pyrolysis.

Pyrolysis is a recycling technique that uses high heat in an oxygen-free environment to decompose plastic waste into its building block components. The heat breaks down the long polymer molecules into shorter segments or monomers, allowing the components to be repolymerized into new, high-quality plastics. This process is also beneficial in the production of fuels such as synthetic gas or oil, proving to be a double-edged sword in the battle against pollution and energy conservation.

Another advanced recycling technology, on a similar vein to pyrolysis, is chemical recycling. In chemical recycling, rather than simply melting and reshaping the plastic, complex chemical processes are used to reduce the plastic back into its base constituent molecules.

The advantage of this process is that it enables the production of plastic items that are of an equivalent quality to their newly manufactured counterparts. This overcomes the primary drawback of traditional recycling wherein the quality of recycled plastic significantly degrades after each cycle.

Solvent-based technology is yet another innovative plastic recycling method. In this technique, a solvent is used to dissolve select types of plastic, after which the plastic-solvent mix is separated from residues and contaminants. The resulting clean,-separated plastic can then be reused to manufacture original-grade plastic products, therefore closing the loop in the plastic life cycle.

Advanced plastic recycling methods are vital in the shift towards a circular economy, where waste is minimally generated, and resources are effectively recycled and reused. Devising efficient, practical, and scalable recycling methods is a key component in this shift, and the latest plastic recycling technologies show immense promise in this aspect.

While the technological aspect of recycling is important, the success of such ventures also hinges on institutional and societal factors. Supportive policy frameworks, sufficient economic incentives, and public awareness, and participation are key elements needed to create a successful plastic recycling program.

Institutional support is crucial in providing the necessary infrastructure for the collection, sorting, and management of plastic waste. Without this infrastructure, even the most advanced recycling technologies would be ineffective.

Additionally, for these advanced recycling technologies to have a sizable impact, they need to be economically viable. This can be achieved through correct pricing of waste generation and treatment, as well as the application of subsidies or tax breaks for recycling activities.

Public participation and awareness are instrumental for the success of recycling programs. Citizens need to be educated about proper waste segregation and disposal. They also need to understand the importance of their role in plastic waste management, from reduction of single-use plastic usage to correct disposal of plastic waste.

While these technologies represent a significant step forward in our approach to managing plastic waste, it's important to remember that they are part of a larger framework of solutions. Reduction and reuse should always be the priority, with recycling used as a solution for unavoidable waste.

Moreover, the adoption and scalability of these advanced plastic recycling technologies are dependent on a variety of factors, including economic viability, market demand for recycled plastic products, local waste management infrastructure, and the level of awareness and understanding in society about plastic recycling.

Despite the challenges, the potential and benefits of these technologies are immense. By incorporating advanced plastic recycling technologies, we can significantly reduce our reliance on virgin plastics, thereby curbing greenhouse emissions and preserving our natural resources. Nevertheless, it's important to simultaneously continue efforts to reduce overall plastic consumption and educate the public on the importance of waste reduction and management.

Industrial Composting

Continuing from our exploration of advanced plastic recycling, let's delve into another important aspect of modern recycling techniques: industrial composting. By definition, industrial composting involves the breakdown of organic waste using microbes under controlled heat and humidity conditions. This process transforms waste into a valuable byproduct known as compost, rich in nutrients and highly beneficial for improving soil quality.

Let's start with the basic premise of industrial composting. The key players in this process are microorganisms that thrive in a ventilated environment rich in carbon and nitrogen, the primary components of organic waste. Given adequate heat and moisture, these microbes rapidly decompose the waste matter into humus, a stable, soil-like material that enriches the soil and promotes plant growth.

Notably, this is a temperature-driven procedure. Desirable composting microbes are thermophilic, implying they operate optimally at higher temperatures, typically between 110°F and 160°F. These conditions speed up decomposition, effectively kill pathogens and weed seeds, and minimize the release of harmful gases such as methane.

Although composting might appear straightforward, achieving optimal conditions on an industrial scale is a technical endeavor. Industrial composting facilities utilize extensive machinery to manage waste, maintain temperature and humidity, and ensure adequate oxygen supply. These methods include windrow composting, in-vessel composting, and aerated static pile composting, each with its own merits.

Windrow composting involves arranging organic waste into long, narrow piles called windrows. Large machines frequently turn these rows to maintain optimal conditions. This is a highly practical method for composting large volumes of waste, such as yard trimmings, food scraps, and even manure.

In-vessel composting, another approach, controls environmental conditions more strictly. Organic material is placed in a drum, silo, or other container where temperature, moisture, and aeration are carefully regulated. This method's advantage is its capacity to rapidly process large amounts of waste within a relatively small footprint.

Aerated static pile composting, on the other hand, requires placing compostable materials on a network of pipes that deliver oxygen to the pile's core. Covered with a layer of insulating material, these piles are not turned like windrows but are allowed to compost naturally. This method is ideal for processing large amounts of homogenous waste material.

Now you might wonder why we should direct so much effort towards industrial composting. To state it simply, this process yields substantial benefits. First, it dramatically reduces the volume of organic waste destined for the landfill, diverting a significant portion of the waste stream towards a more sustainable end. By generating compost, we create a valuable organic product that can replace synthetic fertilizers, enrich soils, reduce water usage, and combat soil pollution.

While industrial composting efforts are robust, it is essential to address some of the challenges associated with it. Contamination of input material, inadequate segregation of compostable waste, and controlling harmful emissions are among the primary issues faced by industrial composting facilities. Intensive monitoring and regulation, coupled with increased public awareness, are critical to address these challenges effectively.

We should also understand that not all waste is created equal when it comes to composting. Species of microbes used, the blend of waste, and the specific composting method can all influence the quality of the final compost. Therefore, there's a distinctive art and science behind achieving a viable, nutrient-rich compost product.

To optimize the gains from industrial composting, one has to consider the role of source separation and contamination reduction. This requires not simply tossing everything into the compost bin but sorting out genuine compostable materials from those that could impair the composting process or reduce the quality of the resultant compost.

There's also the issue of composting bio-plastics, which raises many debates. Some argue that these compostable plastics are beneficial while others suggest they could inhibit the composting process and the quality of the end product. Thus, robust and ongoing research is essential for conclusive judgments.

Industrial composting is indeed a fascinating facet of modern day recycling techniques. It exemplifies the productive intersection of nature and technology where microorganisms do most of the work and machinery aids in setting the stage for them. . With constant improvements in technique and technology, industrial composting provides a promising solution to recycle organic waste while contributing to environmental sustainability.

Recycling of Electronic Waste

In an increasingly digital world, electronic waste, also known as e-waste, is rapidly becoming a significant global challenge. With the incessant production of electronic devices and the fast pace of technological innovation, old devices are frequently discarded, culminating in an immense volume of e-waste.

E-waste consists of all discarded electronic or electrical devices, which, if improperly disposed of, can lead to serious environmental harm. This can be attributed to the release of toxic elements like lead, mercury, and cadmium into the environment. Recognizing this, it's crucial to discuss the recycling of electronic waste as an essential part of modern recycling techniques.

Unlike traditional waste, e-waste requires specific methods for safe and effective recycling due to the presence of hazardous components. These techniques mainly involve manual dismantling, mechanical separation, hydrometallurgical, and biotechnological methods.

Manual dismantling involves taking apart the e-waste to retrieve precious and base metals. This method is labor-intensive but effective in preventing the release of toxic dust into the environment. However, it's important to ensure workers' safety, as they get directly exposed to the harmful components.

Mechanical separation, on the other hand, involves crushing, sorting, and separating e-waste components. It results in the extraction of metal, plastic, and glass, which can be reused or sold. While arguably more efficient and scalable than manual dismantling, it produces waste dust that has to be handled carefully to prevent environmental contamination.

Hydrometallurgical processing exploits the solubility of valuable metals in specific solutions. This method is often used to salvage precious metals like gold, silver, and palladium from e-waste. But it involves the use of hazardous chemicals, which again, need to be handled properly.

Biotechnological methods are considered a greener alternative for e-waste recycling. They involve the use of bacteria, fungi, or other bioagents to extract metals from e-waste. While promising, it requires much more research and development to become a widely used method of e-waste recycling.

Successful recycling of electronic waste also depends heavily on the collection and sorting mechanisms in place. It's vital that consumers are educated about proper e-waste disposal methods. Simply throwing away old electronics with regular trash can lead to harmful environmental consequences. Local communities play a big role in creating drop-off points, while manufacturers can take part in take-back programs.

There's also growing recognition of the need for designing electronics with recycling-friendly features. This implies creating devices that are easy to disassemble and upgrade, reducing the volume of e-waste. Manufacturers' responsibility is crucial here, and some companies have already begun to respond to this call by looking into eco-design strategies.

While these techniques are promising, it's important to acknowledge that the recycling of electronic waste isn't a one-size-fits-all solution. Regulations and practices vary significantly internationally, and low-income countries often lack the necessary infrastructure to safely and efficiently recycle e-waste.

We should also be aware that while recycling provides one pathway to tackle e-waste, it isn't the end of the line. A more holistic approach towards electronic consumption, where recycling is one part of a comprehensive management strategy that includes reducing, reusing, and redesigning electronics, is needed.

Ultimately, the recycling of electronic waste is a multi-faceted process involving technology, policy, education, and individual responsibility. It's an integral part of modern recycling techniques and holds significant potential to alleviate the environmental burden of our contemporary digital lifestyles.

By investing in the technology, regulation, and education needed to support effective e-waste recycling, we may not only prevent environmental harm but also recover valuable materials, create sustainable business opportunities, and move closer to winning the battle against the pressing waste challenge.

Chapter 6:
Individual Responsibility
and Small Changes

In the vast scheme of our planet's predicament, it's easy to feel that individual action may not move the needle much. But it's crucial to remember—big shifts begin with small changes. Let's zoom in to our everyday lives. Each item we consume, each product we purchase, adds a tiny dot to the larger picture of waste and recycling. Our power lies, to a great extent, in how we live our consumer lives. Sure, you can't stop all waste—but you can take steps to reduce personal waste. Picture this: you're at a coffee shop. Order a coffee 'to stay', and you eliminate one disposable coffee cup. Better yet, bring your own reusable mug, and that number grows over time. Reusing and repurposing items can earn more mileage out of many things that we'd often toss without a second thought. Old glass jars can become lovely containers for nuts and raisins. A worn-out t-shirt can start a new life as a grocery bag. It's the Do-It-Yourself Recyclables approach. But where do our products come from, to begin with? It's a question that leads us to conscious consumerism—choosing products not merely based on price or fancy marketing, but looking at their lifecycle. Is it sustainably sourced? Is it manufactured responsibly? Can it be recycled or compose once it's served its purpose? These choices can both limit demand for wasteful products and create markets for responsible, recyclable goods—a small change that can wield a powerful influence in the commercial realm.

Reducing Personal Waste

Perhaps unsurprisingly, the power to institute meaningful change quite literally begins at home. We can all start addressing the issue of waste on a personal level, drastically reducing our environmental footprint by limiting the amount of waste we produce. It is possible to imbue our daily routines and long-established habits with new ways to reduce, reuse, and recycle.

Let's start with composting. Odds are, a significant fraction of your household waste comes from food scraps. By composting these waste products, you can not only reduce the demand for synthetic fertilizers but also the volume of waste you send to the landfill. Additionally, compost enriches the soil in your garden, promoting healthier plant growth.

Now consider your purchasing habits. When buying goods, prefer items that have less packaging or better yet, those that come in recyclable packaging. Single-use plastics have dominated the consumer landscape for far too long, and every choice to avoid them counts.

And it's not merely about what you buy, but how much. Excess consumption leads to waste generation. By adopting a minimalist approach, you can purchase and waste less, leading to substantial waste reduction on a personal level.

Meanwhile, digitalizing our lives can also go a long way in waste reduction. Bank statements, bill receipts, newspapers, magazines, and books can all be accessed electronically, eliminating the need for paper copies. As an added incentive, this not only reduces paper waste but also cuts down on clutter in our personal spaces.

When shopping, make it a habit to carry your cloth bags. This action, though small, can save the number of plastic bags you'd use in a year. Remember, reducing personal waste is as much about embracing good habits as it is about discarding bad ones.

Using rechargeable batteries instead of disposable ones, opting for LED light bulbs, and repairing items instead of replacing them, are among the multitude of small changes that can have a significant impact in the long run. It underscores the notion that it's not just about adopting new practices, but also about reevaluating and reshaping our existing ones.

Eating out, for instance, can lead to a considerable amount of food and packaging waste. Simply choosing to cook at home is an easy way to avoid this. When you do eat out or order take-out, make sure to bring your containers and utensils, to mitigate the waste resulting from disposable dishes, cutlery, and food containers.

Regularly check use-by dates of products in your pantry and plan your meals around items that are soon to expire. This isn't just common sense; it's a natural, effective way to minimize food waste. And don't discard leftovers. Creatively incorporating them into new meals can transform potential waste into culinary delight.

Textile waste is another significant issue. Instead of throwing out your old clothes and shoes, consider donating them, organizing a swap, or turning them into cleaning rags. Similarly, old furniture and appliances can find new homes through donation, exchange, or resale.

Consider buying secondhand items. Not only are they usually cheaper, but they also reduce the demand for new products. By putting less pressure on manufacturers, we can curtail the environmental damages linked to overproduction.

Lastly, but perhaps most importantly, be mindful of your water usage. Wasting water is a luxury modern society can ill-afford. Small changes such as turning off the faucet while brushing your teeth, fixing leaks promptly, and using a dishwasher (yes, it is more water-efficient than hand washing) can greatly contribute to reducing water consumption.

Ultimately, in our mission to reduce personal waste, we must treat what we consider 'waste' altogether differently. Rather than the 'out of sight, out of mind' approach, we need to see waste not as something to be discarded and forgotten, but as a resource that can be harnessed and put to new uses.

Indeed, the path to reducing personal waste is made up of many steps, but even small strides forward represent tangible progress. All it truly takes is a shift in mindset and a commitment to creating a more sustainable planet for future generations.

Reusing, Repurposing, and DIY Recyclables

As we delve further into individual responsibility and small changes to address climate change, one particular strategy stands out: the method of reusing, repurposing, and crafting do-it-yourself (DIY) recyclables. This approach is intertwined with the concept of a circular economy, which we've discussed in the previous chapter. The act of reusing items, giving them a new purpose, and creating homemade recyclables not only reduces waste but also sparks creativity and fosters mindfulness about our consumption habits.

Reusing is perhaps the simplest of the three strategies to adopt. Consider, for example, the packaging in which most of our everyday products are enclosed. Instead of throwing these containers away once the product is finished, we can reuse them for a variety of purposes. For instance, jars and bottles can be used for storage, while cardboard boxes can find a new life as organizers or a canvas for art projects. What makes reusing a powerful tool is that it requires no special technologies or processes and can be done by everyone.

The process of repurposing, on the other hand, takes reusing a step further. It involves modifying an object to perform a function it was not originally intended to perform. An old ladder, for instance, can be transformed into a bookshelf, a windowsill can become a space for growing herbs, or a worn-out sweater can become a cozy blanket for your pet. The potential for repurposing is only limited by one's imagination and creativity.

Do-it-yourself (DIY) recyclables are a more hands-on approach, merging creativity, practicality, and environmental responsibility. They require a higher level of involvement, as they involve creating something new from what is essentially considered 'waste'. For example, you could craft a decorative garden pot from an old plastic bottle or construct a bird feeder out of a milk carton. DIY recyclables are a wonderful way to reduce waste and create something unique and functional at the same time.

One might ask why these practices are vital. Reusing, repurposing, and DIY recyclables help conserve resources by reducing the demand for new products. The less demand there is for new products, the fewer resources are needed to manufacture them. This results in less energy consumption and a decrease in the emission of greenhouse gases.

The impact on reducing plastic pollution alone is significant. By reusing plastics that would ordinarily be discarded, we decrease the amount of plastic waste that ends up in our oceans and landfills. About 8 million metric tons of plastic waste is dumped into our oceans each year, causing untold harm to marine life and ecosystems. Just by hitching a ride on the reuse and repurpose train, we can do our part in cutting down on this alarming figure.

Reducing waste, however, is just the tip of the iceberg of the benefits. By engaging in DIY projects, you can develop new skills, stimulate your creativity, and teach children the importance of resourcefulness, self-reliance, and environmentally conscious behavior. At the communal level, it fosters a culture of creativity and resourcefulness, fostering community pride, and shared respect for the environment.

Furthermore, unlike other environmental solutions, the reuse, repurpose, and DIY approach doesn't need a huge investment or big policy changes to be effective. It fits easily into everyday life and decision-making processes, making it an accessible and engaging way for anyone to contribute to sustainability, regardless of their financial status or societal position.

Another appealing aspect of this approach is that it can be customized to accommodate individual preferences and living situations. For instance, those living in urban settings might focus more on reusing and repurposing due to space constraints, while those in rural surroundings may have more options for DIY recycling projects.

There are countless sources of inspiration and guidance available for anyone interested in these practices. Blogs, YouTube tutorials, DIY workshops, online forums, and social media groups can offer a wealth of ideas for reusing, repurposing, and DIY projects. From turning old wine bottles into beautiful lamps to making benches out of pallets, the possibilities are endless.

Challenging as it may seem to shift the status quo of consumption – buying, using, throwing away – the practices of reusing, repurposing, and DIY provide feasible and enjoyable pathways toward a more sustainable lifestyle. By engaging in these practices, individuals can actively participate in the broader revolution of transitioning from a linear to a circular economy and make their individual mark on mitigating climate change.

The importance of these methods cannot be overstated, not just for the direct impact they have in reducing waste, but also for the awareness and behavioral changes they inspire. Through implementing these practices, people can begin to perceive waste not as garbage, but as resources in the wrong place, nurturing a mentality that can fundamentally change our relationships to consumption, disposal, and conservation.

As we can see, the approach of reusing, repurposing, and DIY recyclables provides an accessible, practical, and captivating gateway to participating in environmental preservation. It reinforces the idea that we all have a role in mitigating climate change and that each small change can make a substantial difference. So let's roll up our sleeves and reinvent the way we perceive and deal with waste!

Conscious Consumerism

As we transition from discussing the primary steps of reducing personal waste and reusing, repurposing and repurposing recyclables, we come upon an essential factor: conscious consumerism. At its core, conscious consumerism involves making purchases in a more mindful and informed manner, with an emphasis on the impact of our choices on the environment and society as a whole. It's a commitment to consume in ways that are ethical, accountable, and sustainable.

Often, the immediate shift isn't about buying more sustainably produced items, but simply buying less. Over-consumption is a significant contributor to climate change and environmental damages, with the production and disposal of goods leading to a bevy of harmful effects. As such, the first principle of conscious consumerism is addressing the question: Do I genuinely need this item?

If the answer is yes, the next step is considering the options. Where does the product come from? Who made it, under what conditions, and with what materials? How lasting is it, and what will happen when its usable life is over? These queries speak to the issues of fair trade, sustainable sourcing, product longevity, and the potential for recycling or composting.

Committing to conscious consumerism involves a preference for goods produced using renewable resources, designed to minimize waste and energy use. In line with the principles of the circular economy, such products are made to be repaired, reused, or recycled, reducing the burden on our planet's resources.

It also means considering the human aspect. Labor rights and equitable trade practices are significant elements of conscious consumerism. Choosing products made by companies that emphasize fair worker treatment, pay appropriate wages, and uphold human dignity helps to promote more equitable economic systems.

Sometimes, it's not just about what you buy but how you buy. Smarter shopping habits like bringing your own bags to the grocery store, choosing loose fruits and vegetables over prepackaged ones, or selecting products with minimal packaging can profoundly decrease the amount of waste generated from our consumption.

Food choices also fall within the realm of conscious consumerism. Opting for local and seasonal produce reduces the carbon footprint associated with the long-distance transport of goods. By favoring organic farming methods, we support practices that are more beneficial to the earth. When it comes to animal products, conscious choices can encourage practices like free-range breeding and responsible fishing, improving animal welfare and sustainability.

Conscious consumerism calls for support to companies with sustainable practices or corporate responsibility initiatives. These could range from waste reduction and energy efficiency steps to contributions to environmental or social causes. Effectively, your purchases become an endorsement of the values such companies uphold.

While it might seem that an individual's choices can't make a significant impact on the grand scale of things, conscious consumerism isn't just about individual actions. It's about creating a collective shift in demand toward sustainable and ethical products. When more consumers demand responsibly produced goods, it exerts pressure on companies to meet that demand, prompting more sustainable practices industry-wide.

Admittedly, navigating the world of conscious consumerism can be complex. Greenwashing, or the practice of making unsubstantiated or misleading claims about a product's environmental benefits, can make it hard to find genuinely sustainable options. To aid in this, there are trusted certifications and labels, such as Fair Trade Certified,

Rainforest Alliance Certified, USDA Organic, and Energy Star, which can guide responsible purchasing decisions.

Additionally, a host of apps and online resources are available to assist in this shift towards conscious consumerism. These assist consumers in identifying the sustainability and ethical credentials of different brands and products, thereby facilitating informed decision-making.

Conscious consumerism, while not a cure-all solution, is a powerful tool in our arsenal against climate change and environmental degradation. By aligning our consumer habits with our environmental values, we can each play a part in fostering a more sustainable, responsible, and equitable world.

As we proceed to the upcoming section about giving new life to used materials, bear in mind that conscious consumerism and recycling go hand in hand. Adopting a conscious consumer mindset helps us choose items with an afterlife, ones that can be reused, recycled, or composted, thereby creating a seamless cycle of responsible consumption and disposal.

Chapter 7:
Giving New Life to Used Materials

In our pursuit of lowered waste and ecological responsibility, there exists a transformative avenue where we turn our waste into a resource, creating a process termed 'upcycling'. Upcycling holds the promise of giving new life to old materials, whether it's plastic, glass, or metals, by creating items of equal or better quality than the original. It's an ingenious way to reel in the seemingly unstoppable tide of waste, turning it around into something valuable again. This chapter also introduces the concept of 'urban mining'. This practice is not about digging into the earth but "mining" our own landscapes of discarded materials, turning e-waste or post-consumer items into valuable resources. This, however, isn't about randomly repurposing everything. It requires a focused effort to identify what materials hold value when recovered and which repurposing methods are most energy-efficient and sustainable. A comprehensive understanding of the materials we use, from their creation to their disposal, is essential in facilitating the reincarnation of materials into useful products, fueling a new sustainable economy built on rejuvenating used materials.

Upcycling and Repurposing

As we continue exploring innovative methods of reducing waste and lessening our environmental impact, let's delve into the intriguing world of upcycling and repurposing. Not only do these methods contribute to a more sustainable planet, but they also encourage creativity and resourcefulness in our everyday lives.

In its simplest form, upcycling means taking something deemed useless or unwanted and turning it into something of greater value. So instead of tossing an item into the recycling bin (or worse, the landfill), you repurpose it into something that's not only functional but even superior in some cases.

Think of it like this: imagine you've received a gift packaged in a beautiful, sturdy box. Instead of throwing the box away after you've opened your gift - which is wasteful - or even recycling it - which consumes energy - you could upcycle it by reusing it as a storage container, maybe decorating it to match your interior decor. The box, then, not only keeps its materiality but also gains a new life and purpose.

Another vivid example that may ring true for many of us: transforming old clothing. Instead of discarding an old T-shirt, it could be upcycled into a bag, a rug, or perhaps even a trendy new top. Upcycling, at its core, is a celebration of creativity, resourcefulness, and environmental consciousness.

The process of repurposing follows a similar philosophy, but it focuses more on reusing rather than enhancing. For instance, one can repurpose a worn-out ladder into a bookshelf without necessarily improving its inherent value. Regardless, through repurposing, we're able to extend the lifespan of objects and reduce waste output.

A key advantage of both upcycling and repurposing is that they demand minimal energy input compared to recycling. Whereas recycling requires considerable energy to break down materials and fashion them into something new, upcycling and repurposing transform the use of things with little to no extra energy required. This difference, while seemingly small, can accumulate impressive environmental benefits when practiced on a large scale.

For instance, imagine the massive energy savings if a significant portion of the population decided to upcycle their drink containers instead of recycling them. While the energy made in recycling is less than the energy required to produce new materials, there's still quite a bit of energy involved in the process. Upcycling, in this case, would save that extra energy while producing an equally functional product.

Now, let's discuss a broader implication. View upcycling and repurposing as not only direct strategies for waste reduction but also key drivers for a paradigm shift in how we perceive and value materials. In a world dominated by a throw-away culture, upcycling and repurposing can provide concrete steps towards building a society where waste is viewed not as garbage, but as a resource.

By encouraging a mindset that values the longevity and potential multifunctionality of objects, we can significantly impact our relationship with consumption. No longer should we view products and material possessions as single-use entities, rather investments that can creatively and resourcefully serve us in numerous ways beyond their initial purpose. This shift can help us break free from the cycle of consumption and discard that's currently prevailing in our culture.

Upcycling and repurposing can also find a place in our economic structures. Allowing companies to find value in would-be waste products could encourage more responsible production practices, cutting down on the unnecessary generation of new materials. This would represent a significant step towards a more circular economy, where the lifecycle of materials is extended, and waste is progressively eliminated.

Furthermore, opportunities for innovation and commerce can arise through upcycling and repurposing. New businesses could potentially flourish, focused on making high-quality, desirable products from waste materials that might otherwise end up in landfills. Such endeavors not only stimulate local economies but also encourage widespread acceptance and practice of these recycling methods.

An important caveat must be mentioned - practicing upcycling or repurposing doesn't mean you can forget about reducing your consumption or recycling your trash. These methods are all part of an overall strategy to lessen our environmental impact and should be used in concert with each other for maximum effect.

Last, but not least, upcycling and repurposing provide a fantastic opportunity to engage communities, especially younger generation in waste reduction strategies. Not only can these practices be fun and engaging, they can also foster a sense of pride and ownership, and impart critical lessons about resourcefulness, creativity, and environmental responsibility.

So the next time you're about to toss something in the trash, take a moment to think about how you might upcycle or repurpose it. You might be surprised at what you can create, and you'll be helping to give new life to used materials at the same time.

Urban Mining: Extracting Value from Waste

As we've discussed earlier, the notion of recycling has expanded far beyond depositing empty soda cans into a blue bin. One prime example of this heightened thought process is the concept of urban mining, a modern method of extracting precious materials and other resources from waste considered as 'urban ores.' Such city-based artifacts include discarded batteries, old electronics, construction rubble, and more.

In essence, urban mining is the process of reclaiming compounds and elements from products, buildings and waste. It is a crucial part of a burgeoning economic model colloquially known as the 'circular economy.' Urban mining plays a pivotal role in ensuring used materials are not thrown away but given new life.

It's easy to view waste as unutilizable, but we need to start seeing it as a resource. Although the name suggests large-scale operations typical in traditional mining, urban mining can occur on any level, from individual households to entire cities. Feel inspired by the thought that your old laptop could be a gold mine, literally. You can find up to 800 times more gold in a ton of discarded mobile phones than in a ton of gold ore.

So, how does one conduct urban mining, you might ask? To start, it requires an innovative approach to standard waste management procedures. Instead of discarding trash directly into a landfill, we sort, shred and process it to extract valuable components. Next, refinement processes are employed to separate these elements, transforming them into viable commodities that drive the economy. The fruition of these processes can be exemplified by extracting copper from discarded electrical wires or harvesting asphalt from old roads.

Urban mining's economic advantage lies in its application to electronic waste, the fastest growing waste stream in the world. With an annual global generation of nearly 50 million tons, electronic waste, fondly named e-waste, is a gold mine for those ready to dig. We must recognize the economic potential of this technology-driven waste deluge. From e-waste, we can reclaim valuable materials like gold, silver, copper, and rare earth elements that would otherwise require intensive mining to obtain.

Not only does urban mining divert waste from landfills, thereby reducing landfill space and the subsequent generation of harmful greenhouse gases, but it also reduces the energy and resources required for traditional mining activities. This reduction subsequently decreases our carbon emissions and our overreliance on finite natural resources.

Urban mining also promotes better health and safety standards. Traditional mining often involves dangerous environments and toxic chemicals, placing workers at risk. On the other hand, urban mining can be mechanized and controlled, thereby ensuring safer workspaces while maintaining environmentally friendly practices.

However, there are definite challenges to urban mining that cannot be overlooked. These include difficulties in collecting and sorting waste, lack of public awareness, and technological constraints. But, as you'll come to understand, such obstacles are far from insurmountable.

Most importantly, we should emphasize that urban mining is not a standalone solution for our waste dilemma but a piece of the puzzle. It must be not only applied but also supported by other practices such as efficient recycling systems, conscious consumerism and legislative measures. Collectively, these efforts converge to produce a more effective, sustainable waste management strategy.

While urban mining may be a relatively new concept, it's essential to understand its potential impact in our fight against climate change. New pioneers are seeing urban landscapes not as mere buildings and infrastructure but as treasure troves of valuable resources. This new viewpoint is the paradigm shift needed to transition from our linear economy to a circular one.

Through urban mining, we can transform cities into never-ending recirculating loops of resources. This makes waste a misnomer, as nothing created or consumed in an urban environment is ever truly wasted. Instead, the end of one product's life merely indicates the start of another's.

Tomorrow's miners won't need to delve deep beneath the earth to extract precious metals. Instead, they might be scouring your old computers or broken household items. Knowingly or unknowingly, we are part of an urban mine. Let's match the scale of our consumption with the scope of our recycling efforts to help pave the way for a future where waste is a myth.

To conclude, creating value from our waste through urban mining isn't just a fascinating concept, it's a necessity. It offers a viable solution to our resource-driven society, which is currently exploiting our planet's finite resources at an unsustainable rate. The promise of urban mining isn't merely to maintain our lifestyles but to improve them and ensure our planet's longevity.

Chapter 8:
Go Beyond Recycling:
Reclaiming Our Planet

It'll take more than recycling to bring back our once-vibrant planet. We need to re-establish a symbiotic relationship with the environment while configuring human consumption habits in harmony with nature. We should look for innovative ways to restore natural habitats, which have been devastated due to rapid urbanization and unfettered extraction of resources. Repopulating these spaces with indigenous plants and animals fosters biodiversity and revitalizes ecosystems. However, this might not be enough unless we tackle the energy crisis. Climate change means we can't rely on non-renewable resources. It's crucial to shift our focus towards renewable energy sources like wind, solar, and hydro. Let's not stop at harnessing these for generating electricity, but also strive to fuel our automobiles and heat our homes sustainably. Lastly, it's critical to examine our dietary habits and agricultural practices. By endorsing organic and regenerative agriculture we not just nourish ourselves but also our soil, reducing the dependency on chemical fertilizers and promoting the health of Earth's bee population. Embarking on these actions will take us a step closer to not merely recycling, but truly reclaiming our planet.

Restoring Natural Habitats

As we continue our exploration of methods to salvage and rehabilitate our world, it's important to talk about restoring natural habitats. Disruption to natural habitats due to human activities like deforestation, urbanization, and climate change is devastating. It's not enough to recycle, we need to actively work towards regenerating biodiverse habitats that make our planet healthy.

Our role is crucial. Native ecosystems need to be restored, as they're the bedrock of our planet's biodiversity, capturing and storing vast amounts of carbon dioxide and providing a refuge for wildlife populations. Their preservation and restoration may be one of the most effective tools against climate change at the ready, and growing awareness of this fact has triggered a global push for habitat restoration.

Forest and wetland restoration is valuable. Opting for organic, deforestation-free products is a personal choice we can make to fight against habitat loss. We can also contribute by partaking in restoration initiatives or assisting organizations that focus on habitat restoration, as well as advocating for forest conservation policies at the government level.

Urban areas aren't exempt from the need for restoration. The introduction of green spaces in the heart of the concrete jungle can create micro-habitats for local flora and fauna, simultaneously improving the urban environment's air quality and aesthetics. These initiatives spearhead the movement for more sustainable cities, and we all have a role to play.

Every small action counts, like planting native species in your backyard or community gardens to promote local biodiversity or participating in community clean-up campaigns of local habitats from litter and pollution. Creating insect-friendly gardens or installing bird-boxes in our homes can support smaller species that form the base of numerous food chains.

Another crucial aspect of restoration focuses on protecting and rebuilding our oceans. Marine habitats, such as coral reefs, seagrass meadows, and mangrove forests, are incredibly efficient at trapping carbon but are currently being exploited and destroyed at an alarming rate. Various initiatives work to rehabilitate damaged marine areas and stop harmful activities, such as overfishing and pollution, that further degrade these habitats.

We must also strive to diminish harmful substances draining into our oceans. This can be achieved with more efficient waste management, mitigating runoff from agriculture, and reducing our plastic usage to lower the amount of debris entering marine ecosystems.

Restoring natural habitats also involves the heartlands, our prairies and savannas. These ecosystems are rich in biodiversity and work as efficient carbon sinks. Supporting projects that expand these expansive environments and combat desertification processes are invaluable.

Returning to more traditional agricultural practices can have significant benefits too. Introducing agroforestry, a system that integrates trees into farmland, can enhance biodiversity, combat erosion, and improve water quality, thus creating more resilient ecosystems.

In all of these initiatives, it's vital for us to seek to work in harmony with the natural world, rather than against it, a practice known as rewilding. This means looking at ways to restore wilderness regions and protect them, allowing nature to take its course and encouraging wildlife back into these areas.

However, it's not just about the physical restoration of habitats, but also about how we perceive our relationship with the natural world. Considering nature as a partner rather than a provider promotes a mindset shift that's key to fostering successful and long-lasting restoration projects.

The journey towards restoring natural habitats is multifaceted, and each step taken interlinks with another. Whether it's small steps such as planting a tree, supporting sustainable businesses, or campaigning for stronger governmental policies, it's no doubt that every action helps rebalance our ecosystem.

To truly reclaim our planet, we can't just focus on reducing our carbon output and improving recycling. We need to address the damage and disruptions caused by our previous actions and work towards restoring the natural habitats that are the essence of life on this planet.

By putting into practice the principles and actions highlighted in this section, one thing becomes clear: every one of us has the power to aid in restoring natural habitats and the responsibility to shape a healthier planet. The journey may be long and non-linear, but it is one we must embark on with determination, grit, and unwavering optimism.

Support Renewable Energy Sources

Transitioning to renewable energy is a pivotal part of our evolving story of reclaiming our planet. As we all know, the earth can't keep up with the demand we are placing on it. Our consumption patterns create a perilous dance between the need for energy and the irreversible damage we inflict on our environment. Solar, wind, hydroelectric, geothermal, and biomass sources of energy have become beacons of hope in this dance, setting the tempo for a sustainable and viable future.

First, what do we mean by renewable energy? At the crux, it's energy derived from resources that are naturally replenishing, such as sunlight, wind, tides, and geothermal heat, which are sustainable and do not pollute our environment. We have technologies to harness these types of energy, but the effective distribution and accessibility still pose a challenge.

Solar energy is perhaps the most recognized form of renewable energy and for good reason too. The sun provides an abundant and consistent source of energy- it's really the perfect power plant! Not only is solar energy clean and inexhaustible, but it's also becoming less expensive with each passing year. It only accounts for about 1.7% of the United States' electricity, which is a number we can definitely increase. Increased efficiency in powering homes through solar panels and new innovations like solar windows can significantly reduce our reliance on fossil fuels.

Wind energy is another potent choice that is rapidly growing in adoption. Wind turbines, whose design has improved significantly in recent years, harness natural wind energy to generate electricity. Today, about 8.4% of US power generation comes from wind energy. To further fuel this growth, offshore wind farms must be implemented, which have additional potential due to consistent, high-speed winds in these areas. This could increase the consistency of power generation and it also avoids the common criticism of spoiling visual landscapes.

Hydroelectric power, generated by water stored in dams or flowing rivers, forms a significant part of the power matrix for many countries, including the United States, where it constitutes nearly 7% of total electricity production. Upgrades in turbine technologies are making hydropower a more efficient and less ecologically destructive energy source. Also, the concept of micro-hydros, which are small, low-impact systems designed to generate power for individual homes or small communities, is garnering interest.

Geothermal energy, born deep within the earth, leverages the planet's own heat. Still a relatively untapped resource, geothermal power plants are capable of producing electricity continuously, making it a dependable source of power. Enhanced geothermal systems can be developed in warmer locations, removing the need for convenient natural reservoirs, opening up geothermal power to broader usage. Just imagine a world where houses are heated in winter by the earth's natural warmth!

Biomass power, generated from organic material, has been used throughout human history. Modern biomass power plants are carbon-neutral, as the carbon dioxide released is reabsorbed by trees and plants. Truly sustainable and part of a circular economy, biomass power utilization awaits much-needed innovation and breakthroughs.

Transitioning to renewable isn't just about the environment, it's also about social and economic benefits. More demand for renewable energy products means more jobs in manufacturing, installation, maintenance, and research and development. This can drive significant economic growth.

However, the shift happens not just at an industrial or governmental level, but also with individuals. As consumers, we have tremendous power to direct the momentum of this transition. For instance, opting for utility companies that source a significant proportion of their energy from renewables, or installing solar panels and wind turbines where possible, can represent important contributions.

Among all the options, energy efficiency serves as an often-overlooked form of sustainable energy. Shutting off lights, unplugging unused appliances, and using energy-saving devices - these don't necessarily require massive changes, just tweaks in our everyday habits.

Supporting the development and adoption of renewable energy is even more than a choice, it is a collective responsibility. It's no longer just about being 'cool' and 'green'. It's about survival, human rights, social justice, and it's about ensuring a viable future for coming generations.

So, let's embrace renewable energy in our individual capacities and demand it from our policy makers. Let's not shy away from asking tough questions from corporations about their energy sourcing. In this way, we will accelerate the shift from polluting fossil fuels to clean, renewable sources of energy. This is not something we merely should do, but something we absolutely must do. Our planet depends on it.

Organic and Regenerative Agriculture

Veering away from focusing solely on waste management, we now turn our attention to a powerful strategy that has the potential to overturn the hazardous impacts of climate change and rejuvenate our planet: Organic and Regenerative Agriculture. While our modern agricultural systems have been aimed towards large-scale production, they have unfortunately resulted in significant damage to our environment.

Organic agriculture leans towards farming techniques that avoid the use of artificial chemicals, pesticides, and genetically modified organisms. In pursuing this ecologically tuned method, not only can we promote biodiversity and improve soil health, but we can also contribute to lowering greenhouse gas emissions.

Regenerative agriculture, on the other hand, focuses on rebuilding organic matter in the soil and restoring degraded soil biodiversity. This approach often includes techniques such as crop rotation, composting, no-tillage, and diversified cropping systems. The end goal is to create a system that not only sustains but enhances our agricultural resources.

Such practices can't be understated in their importance. They play a crucial role in countering some of the most detrimental effects of climate change. To explain why, we must understand that our soil contains a vast amount of carbon. When managed poorly, this carbon can be released into the atmosphere, contributing to global warming. However, when managed well, as in the case of regenerative farming, the soil's carbon can be preserved and additional carbon can be drawn down from the atmosphere—a process known as carbon sequestration.

Furthermore, organic and regenerative agricultural practices help restore the nutrient balance in our soil. Increased microbial activity contributes to nutrient cycling, promoting the availability of nutrients to crops. This leads to healthier and more nutritious harvests, in turn supporting human health and well-being.

This cultivation method also supports the surrounding ecosystem. The use of cover crops and no-till practices reduces soil erosion, preserves water resources, and increases biodiversity. Implementing polycultures and agroforestry can provide habitat for a wide variety of species, creating a balanced and resilient ecosystem.

Interestingly, the adoption and implementation of organic and regenerative agriculture is not without its challenges. Some critics argue that these practices can't meet the massive food production demands of our growing global population. However, studies suggest that if done right, using a combination of regenerative practices can indeed achieve high levels of productivity.

Adopting such agricultural practices requires a considerable shift in mindset. We must recognize the interconnectedness of our actions and the environment. Instead of viewing farming as a process to extract as much yield as possible, we need to treat agriculture as a method to nurture and give back to Earth.

Transitioning to regenerative agriculture also requires the willingness to learn new techniques and to invest time and resources into sustainable farming practices. Community-based education initiatives and classes can be instrumental in supporting farmers in this transition.

As consumers, our choices can also impact the prevalence of organic and regenerative agriculture. We can choose to buy products from companies that are transparent about their supply chain and that employ sustainable and regenerative farming practices. The rise in farmer's markets and direct-to-consumer sales models present opportunities for consumers to support local, small-scale farmers who are using these practices.

Moreover, developing and implementing policies that promote organic and regenerative agriculture plays an important role. Government incentives, grants, and resources can encourage farmers to transition towards these practices, thereby accelerating the agricultural shift towards sustainability.

In conclusion, the role of organic and regenerative agriculture is not limited to producing high-quality food. These systems offer an opportunity to reclaim our ailing planet by enriching our soils, enhancing biodiversity, sequestering carbon, and supporting resilient local economies. As we strive to combat climate change and promote sustainability, let's not forget the traditional wisdom embedded in our soils and keep the conversation growing around organic and regenerative agriculture.

Chapter 9:
Policy and Action: The Way Forward

It's clear we've reached an environmental crossroads. The success of our recycling efforts lies not just in individual actions, but in collective policy decisions as well. Government policies on climate change and recycling can shape major systematic shifts. We've already seen the transformative impact of regulations around single-use plastics in some regions, and the burgeoning popularity of carbon pricing schemes rewarding companies for greener practices. These policy changes demonstrate the necessary interaction between economy and ecology. In addition, corporations are stepping up to drive sustainable change. From tech giants harnessing green energy to retailers ditching non-recyclable packaging, they're demonstrating the potential of a circular economy. This doesn't absolve them of past environmental missteps, but it shows that we can reward businesses who choose to be part of the solution. Lastly, regular citizens have a responsibility to maintain pressure through activism, lobbying for more stringent environmental protections and voting with our wallets. We must remember - the future isn't something that just happens to us, it's something we're actively creating. The way forward, thus, is paved with policy, action, and the unyielding belief that we can make a difference.

Government Policies on Climate Change and Recycling

The interplay between government policy and environmental issues like climate change and recycling is unequivocal. Policy is a critical tool in countering the problems posed by climate change and insufficient recycling practices. Government intervention in the form of well-structured policies, regulations, and initiatives is a necessity in our journey towards a sustainable planet.

America's stance towards climate change has witnessed some momentous shifts through recent years. The U.S., historically one of the world's largest carbon emitters, initially had a relatively tepid response to the global climate crisis. However, the latter years exhibited a more proactive approach in mitigating climate change through the adoption of several key policies and agreements.

One such pivotal move was the adoption of the Paris Agreement under the United Nations Framework Convention on Climate Change (UNFCCC). Central to this agreement is a commitment by developed countries to make "finance flows consistent with a pathway towards low greenhouse gas emissions and climate-resilient development." In simpler terms, it's about investing where it counts to curb the looming impacts of climate change.

Coming closer home, domestic policies like the Clean Power Plan have also emerged. This ambitious plan aimed to reduce carbon dioxide emissions from electrical power generation by 32 percent by 2025. Although the policy has faced political roadblocks and legal battles, it stands as a testament to the government's recognition of the climate change problem.

In tandem with policies targeting climate change, government initiatives towards effective recycling are equally essential. Waste management and recycling are often regulated at the local level with differing policies across states. However, federal agencies can influence these practices through grants, guidelines, and regulations.

The Resource Conservation and Recovery Act (RCRA), for instance, grants the Environmental Protection Agency (EPA) the authority to control hazardous waste from 'cradle-to-grave'. This includes the generation, transportation, treatment, storage, and disposal of hazardous waste. Additionally, the RCRA gives the EPA the responsibility to establish guidelines for recycling and waste treatment.

Another policy worth mentioning is the Comprehensive Environmental Response, Compensation, and Liability Act (CERCLA). Commonly known as Superfund, this program aims to clean up sites contaminated with hazardous substances and pollutants. These policies signify the government's commitment to reduce waste, promoting recycling, and ensuring a safer and cleaner environment.

As encouraging as these policies are, there's significant room for improvement. Substantial gaps persist in the regulations surrounding hard-to-recycle materials such as e-waste and plastic. Furthermore, recycling rates in the U.S. are still relatively low compared to other developed nations.

A broader policy approach incorporating extended producer responsibility could be the key. This approach places an obligation on producers to manage the environmental impact of their products, even after consumer use. By incorporating the full lifecycle costs of products, this policy could drive the transition to a more circular economy.

There is also a need to develop policies that encourage the use of recycled materials over virgin ones. These could involve providing financial incentives to manufacturers for incorporating recycled components in their products, which would help stoke demand for recycled materials and stimulate the recycling industry.

In conclusion, while current government policies have made some positive strides in combating climate change and promoting recycling, the scale of the challenge ahead calls for a high level of creativity, persistence, and fortitude. More comprehensive and innovative policies, such as those promoting extended producer responsibility and the use of recycled materials, can have a transformative impact on our waste generation and disposal behaviors.

Adopting these policies requires a significant commitment from all stakeholders, including policymakers, businesses, and consumers. However, the potential rewards - a healthier planet, more sustainable consumption practices, and a thriving 'green' economy - make this challenging endeavor unequivocally worthwhile.

Through sustainable and globally-minded government policies, we can inspire businesses to innovate in recycling processes, compel individuals to implement sustainable practices, and motivate the world to act against the pressing problem of climate change. Indeed, with the right governmental policies to steer societal actions, we can collectively work towards achieving a sustainable, climate-resilient future.

Companies Making a Difference

The corporate world can no longer ignore its environmental responsibilities. There are many notable companies leading the change by developing innovative policies and initiatives. This chapter will focus on pioneering companies that are helping shape a more sustainable world.

Firstly, we must mention Patagonia. This outdoor gear company has pushed boundaries since its establishment in 1973. Patagonia is not just manufacturing sustainable goods, but also taking considerable steps to minimize its environmental impact, like promoting the repair and recycling of its products. Additionally, the company pledges at least 1% of sales to environmental groups through its self-imposed "earth tax".

Interface, a global manufacturer of commercial and residential carpets and resilient flooring, is another outstanding example. They've created unique programs to reduce their reliance on virgin materials and cut carbon emissions. Their ambitious climate goal, "Mission Zero", intends to eliminate any negative effect the company might have on the environment by 2020.

We also have Tesla, a company associated much with a unique perspective on electric vehicles (EVs), but their mission is broader than that. Tesla aims to 'accelerate the advent of sustainable transport', transforming the entire automobile industry in the process. Their renewable energy products, like the solar roof and Powerwall, show that they're committed to creating sustainable energy solutions.

We can't exclude EILEEN FISHER from the line-up. A leader in the fashion industry, EILEEN FISHER pursues a policy of supply chain transparency and invests in initiatives to recycle used garments into new products. Their "Waste No More" campaign embodies their commitment to zero waste.

Notably, Google's commitment to renewable energy deserves our attention. Google has been carbon-neutral since 2007, and in 2017, they reached 100% renewable energy for all their operations. Moreover, the company's AI improves the efficiency of wind and solar energy, proving that a tech company can have a positive environmental impact.

Even companies who are traditionally seen as part of the problem are making strides towards becoming part of the solution. Take Coca-Cola, for example, who announced their "World Without Waste" initiative, vowing to recycle a bottle or can for every one they sell by 2030.

IBM, a tech giant, has been prioritizing sustainability for decades. Their focus includes reducing CO2 emissions, conserving energy, and facilitating responsible water management. Also, they've created a comprehensive recycling program for electronic waste, setting an example for the rest of the tech industry.

Salesforce has also surfaced as a corporate leader for the environment. This customer relationship management (CRM) company is committed to reaching 100% renewable energy use. Salesforce also cultivates a culture of sustainability among employees, encouraging them to take climate action via their 'Planet Salesforce' initiative.

General Motors (GM), an unlikely suspect, has begun to embrace sustainability. GM operates landfill-free manufacturing facilities, promotes resource preservation and is making solid strides towards electric and zero-emissions vehicles. They aim at setting a new standard in the automobile industry.

Nike has been known to set trends in the athletic department, and sustainability is no different. Nike focuses on creating products that minimize waste and use environmentally preferable materials through their 'Move to Zero' initiative. They're also advocating for policy changes that combat climate change.

Lastly, Unilever can't go unnoticed. Their Sustainable Living Plan is designed to decouple their environmental footprint from their growth. Unilever aims at halving their environmental impact while doubling their size. Moreover, they're investing in renewable energy, water conservation, and waste reduction.

These companies realize that being environmentally conscious does not only make moral sense, it makes business sense too. They're showing that profitability and sustainability can and should coexist.

By integrating environmentally friendly practices into their business models, these corporations are proving that it's possible to adjust course and reduce our collective impact on the planet. They're challenging the status quo, investing in our planet, and paving the way for a sustainable future.

While it's often easy to criticize big corporations for their environmental impact, it's also important to recognize and support the companies that are making efforts to change. They've taken up the mantle of corporate social responsibility, and it's time for others to follow suit.

When we think of making a difference, we often think of the actions individuals can take. But companies have a significant role to play in this as well. Through innovative business practices and a commitment to environmental responsibility, these companies are helping pave the way toward a more sustainable future.

Activism and the Responsibility of Citizen

In our discussion thus far, we've explored the numerous issues surrounding our planet's current state, the prevalence and impact of waste, our consumption habits, and the modernization of recycling practices. Now that we have an understanding of these topics, it's crucial to harness our newfound knowledge and transform it into action. As we step forward on this journey, we'll delve into the sphere of activism and the integral role of individual responsibility.

Activism isn't exclusively about attending rallies or protests; it's also about educating oneself, spreading awareness, changing everyday practices, and contributing to the larger movement for environmental sustainability. The quest for a sustainable future isn't a solitary endeavor. It involves each individual realizing their potential to influence change, regardless of how big or small these changes may seem.

We can't underestimate the power of individual action. Collectively, our seemingly inconsequential actions can lead to significant positive changes. Each time we model sustainable behaviors, whether by recycling correctly, reducing waste output or engaging in conscious consumerism, we're performing an act of activism. We're breaking away from the unsustainable norm and paving the way for a greener future.

Yet, individual acts of environmental stewardship alone, while beneficial and necessary, won't solve the climate crisis. We must remember that structural changes are vital too. Citizens can amplify their impact by advocating for broader systemic changes within governmental policies, corporate responsibility, and societal norms surrounding consumption and waste management.

Another potent tool in the activist's arsenal is the power of the vote. Each voting decision, local or national, reflects an individual's commitment to environmental protection. Voting for leaders who unequivocally invest in eco-friendly policies, push for green infrastructure, and acknowledge the urgency of climate change is a form of activism that holds far-reaching implications.

Beyond political involvement, activism encompasses consumer power. Consumers can sway corporate behaviors through their purchasing decisions. Supporting businesses that prioritize sustainability over short-term profit sends a strong message to the market about what consumers value. From the food we eat, the clothing we wear, to the energy we consume, every purchase has the potential to drive change.

Advocacy and activism also encompass digital spaces. In today's technology-driven world, the Internet provides a valuable platform for education and mobilization. Sharing accurate information, signing online petitions, or joining digital campaigns are all ways to boost climate awareness and Recycling 2.0 worldwide. The reach and influence of digital activism should not be underestimated.

Community involvement is another essential aspect of citizen responsibility. Local community events such as clean-ups, recycling drives, and educational workshops can foster a spirit of collaboration and shared responsibility. The community network becomes a nurturing ground for collective action and mutual encouragement.

Of course, any mention of environmental activism would be incomplete without acknowledging the young voices leading the charge. The youth's involvement in climate advocacy groups has brought renewed energy and urgency to the environmental movement. They carry the baton for future generations and act as compelling reminders of why this fight is crucial.

Education plays a key role in fostering responsible citizen activism. A comprehensive understanding of climate change, waste management, the circular economy, and modern recycling equips individuals to make informed decisions and contribute to public discourse. As we enhance our knowledge and understanding, we can stand as stronger advocates for the planet.

Success in tackling environmental issues won't be swift or easy. However, the investment in our planet's future is well worth the effort. Restoring the Earth requires an all-hands-on-deck approach, with every citizen stepping up to the mark.

Activism is not a buzzword to be trendy or popular. It's the active participation in the long-term fight for our planet's health. At its core, it involves a willingness to change, to learn, and to act. It's about recognizing our interdependence on the Earth and consciously deciding to walk a more sustainable path.

As we close this chapter on activism and the responsibility of citizens, may we remember: the goal isn't to fight against the Earth, but rather to live in harmony with it, ensuring a sustainable future for all. The call to combat climate change and waste is a call for us, as individuals, to step up, do our part, and bring about the changes we need to see.

As we open the next chapter, we'll explore a vision for a future where we've not only revitalized recycling but also redefined our relationship with the environment. The actions of today will echo in the generations of tomorrow, and it's up to us to secure a flourishing planet for ourselves and those who follow after us.

Chapter 10:
Final Thoughts: A Progressive Planet
for Future Generations

As we look ahead, it's evident that crafting a sustainable future isn't an individual venture, but a collective responsibility shared by each of us, young and old. One of the key factors in nurturing a resilient planet for future generations is mitigating climate change and this we can't achieve without a sustained commitment to novel, efficient recycling techniques. Modern recycling practices provide a beacon of hope, offering innovative approaches like urban mining and industrial composting, that not only reduce waste but repurpose it, invariably contributing to the health of our planet. This, combined with policy and action from governments, activism and corporate responsibility, ensures that recycling takes center-stage as a solution, not merely an amelioration. As we turn our gaze to the future, it's clear that the possibilities for recycling are as boundless as our collective resolve and ingenuity. As we embrace these solutions, we pass on a legacy of health, sustainability, and respect, teaching generations to come the enduring value of protecting and conserving our shared home. On this note, we close with the hope that the seeds sown today grow into a green, thriving planet for all those who follow.

Building a Sustainable Future

As we navigate the path forward, the key realization worth focusing on is that each of us has a role in constructing a sustainable future - a future where there's harmony between the planet's resources and our consumption patterns. Our monumental progress in modern recycling techniques and circular economy practices, as well as our commitment to minimize waste, are all stepping stones in this direction.

A vital component of building a sustainable future is shifting our perception of waste completely. Instead of viewing it as a problem, we need to see it as a resource in itself. Alongside minimizing waste generation, we must optimize the utilization of the waste we do produce.

Just like nature, where nothing goes to waste, we too should strive for a zero-waste approach. From industrial composting of food waste to advanced recycling technology for plastic, we've made significant strides toward turning waste into wealth. Further, through urban mining, we're extracting valuable resources from our waste.

But, recycling is just one part of the story. For true sustainability, we need to facilitate the restoration of natural habitats, support renewable energy sources, and promote organic and regenerative agriculture. In other words, cultivating a culture of conscious consumerism.

Technology isn't the only answer; we need system-wide transformations. An inclusive legislative environment that encourages innovation and enforces responsibility is paramount. Companies need to be accountable for their waste, and those taking proactive steps towards sustainability should be recognized.

There's power in collective action. Small changes in our own lives when scaled up can lead to significant impacts. Simple habits such as reducing personal waste, reusing items, and choosing products made from recycled materials are individual steps we can take.

But, we must go beyond individual actions and become catalysts for change within our societies. Advocacy and civic involvement are crucial for driving large-scale transformations. From local community recycling projects to global climate marches, we can become forces for change.

We must also bring focus to sustainable education. By equipping future generations with knowledge of climate science and waste management, we're essentially powering them to innovative solutions. Learning about the implications of plastic pollution or the merits of a circular economy from a young age can foster an attitude of responsibility.

Our infrastructures need to be designed with the end in mind. That is, they should focus on the ease of repurposing or recycling at the end of their useful life. Whether it's the construction of green buildings or the design of electronics, sustainability should be inherent in the lifecycle.

As we forge ahead in constructing a sustainable future, we must always remember that it is not a destination but a journey. A journey that requires constant learning, adjusting, and innovating. And in this journey, every step we take, no matter how small, leads us towards a healthier, more sustainable planet.

Whether it's through recycling, conscious consumerism, policy advocacy, or individual action, we all can contribute to building this sustainable future. Our actions today are effectively the building blocks for the kind of world we want to leave behind for future generations. Through our collective efforts, we can ensure that our progress does not come at the detriment of our planet.

Indeed, we are the gatekeepers of our planet. The future sustainability of our world is in our hands, and each step we take, each decision, and each action leads us closer to or away from that sustainable future. And while the task may seem daunting, let it not deter us from striving for what we know is possible.

In concluding, building a sustainable future is more than just an environmental imperative; it is a shared vision that drives towards a world where people, profit, and planet are in harmony. It's a journey we all must embark on if we're to ensure a resilient future for ourselves and generations to come. It's now up to us to turn this vision into reality.

And so, we look to the future, not with despair or helplessness, but with hope, courage, and the determination to create a healthier, more sustainable world for ourselves, and above all, for future generations.

Climate Change Mitigation: The Role of Recycling

In the context of the escalating climate crisis, the significance of recycling cannot be overstated. Recycling serves as one of the core strategies to combat the dire impacts of climate change. This chapter aims to underscore the role of recycling in mitigating climate change and fostering a sustainable future.

At its core, recycling contributes significantly to reducing greenhouse gas emissions, a leading cause of climate change. Material production, which includes extraction, refinement, transportation, and manufacturing, accounts for a notable portion of global greenhouse gas emissions. By giving waste another life, we're cutting down on the need to create new materials and hence, offsetting the emissions associated with the production process.

Consider the life cycle of an aluminum can. This process begins with mining the aluminum ore, refining it into pure aluminum, manufacturing the cans, and shipping them off to be filled and sold. The end of this life cycle typically involves the can being discarded into a landfill. However, if the can is recycled, it can bypass much of this energy-intensive process. By minimizing the need for new aluminum, recycling drastically reduces greenhouse gas emissions.

Moreover, recycling plays a crucial part in conserving our natural resources. The practice of extracting raw materials is largely detrimental as it disrupts ecosystems, causing deforestation and loss of biodiversity. Recycling helps in cutting down the extraction and depletion of these resources, thereby safeguarding our biodiversity and ecosystem stability.

In the larger picture, recycling acts as a buffer against deforestation. By recycling paper and cardboard, we can significantly reduce the demand for virgin pulp, which should lessen the pressure to cut down forests. Given the role of trees in sequestering carbon, preserving our forests is formidable for mitigating climate change.

Though often overlooked, the role of recycling in water conservation is also indispensable. Manufacturing processes often require substantial amounts of water. By reducing the need for virgin materials through recycling, we lessen the stress on our already strained water resources.

Recycling also plays a critical role in reducing waste that ends up in our landfills. Overfilled landfills are not only environmental risks, but they also contribute to climate change. When organic materials in these sites decompose, they produce methane, a greenhouse gas significantly more potent than carbon dioxide. Therefore, by boosting recycling, we can cut down this methane production.

It's also crucial to realize that recycling innovations, such as advanced plastic recycling and electronic e-waste recycling, are invaluable in tackling climate change. These innovative practices are essential to handling today's incredibly diverse waste streams and can deliver significant greenhouse gas savings.

The fact that the world's waste generation is projected to increase by a whopping 70% on current levels by 2050 according to the World Bank necessitates the need for robust recycling systems. With such increased waste, recycling services could contribute a great deal to climate change mitigation. Revolutionizing the way we think about and manage our waste couldn't be more critical for halting catastrophic climate change.

While recycling plays a pivotal role, it's important to remember that it's just one part of the solution. Comprehensive climate change mitigation will necessitate a broad range of efforts including consuming consciously, reducing where we can, reusing old items, supporting renewable energy, advocating for environmentally friendly policies, and stimulating the transition from a linear to a circular economy. Recycling can't solve climate change on its own, but it can contribute significantly when combined with these other measures.

The role of recycling in climate change mitigation illustrates how interconnected our actions and their impacts on the earth truly are. Addressing the climate challenge necessitates a collective effort and thoughtful consideration of our everyday practices. Our individual and collective actions can, and will, make a difference.

So, let's embrace recycling not just as an obligation, but as a commitment to sustainably coexist on this magnificent planet. Climate change might be an imposing challenge, but with proactive measures, including improved recycling systems, we can march towards a healthier, greener world.

As we continue to innovate and refine our waste management practices, we must remember the critical role recycling serves in our larger journey as stewards of the earth. A comprehensive approach to recycling, twinned with an eye towards sustainability in every aspect of our lives, can provide us with the tools and the vision to leave a sustainable world to future generations.

The Future of Recycling: Predictions and Possibilities

As we march forward in time, confronting the monumental issue of climate change, one can't help but wonder - what does the future of recycling look like? Pondering this question is critically important, as it holds in it the blueprint for a sustainable, environmentally healthy planet. As we've already delved into existing recycling methods and their impact on the planet, let's now turn our gaze toward the horizon and explore some predictions and potentialities for recycling's future.

Firstly, we have technological innovation. Under the glaring spotlight of climate crisis, we're seeing a global turning point in favor of sustainability. Predictably, technology is rushing to meet this demand, and strides are being made toward the development of more advanced, efficient recycling techniques. Particularly, in areas where current methods fall short, such as plastic and electronic waste recycling, innovation is necessary and imminent.

Given the rapid accumulation of electronic waste and its hazardous impact, we predict an uptick in research and development around efficient electronic waste recycling. We foresee a future where recycling electronic waste isn't just an afterthought, but an integral part of product design and lifecycle. Evolved recycling technology can potentially create closed-loop systems where electronic devices are manufactured, used, and then recycled seamlessly into raw material for new devices.

The same vision of a closed-loop system is projected for plastic recycling. Plastic's durability, which contributes to its waste issue, can also be seen as an opportunity. Innovative recycling technologies could enable us to continuously recycle plastics without significant quality loss, turning this notorious pollutant into a poster-child for the circular economy.

Additionally, we anticipate a future where recycling isn't just centralized but localized too. Small-scale, home recycling appliances could be a reality, reducing waste transportation emissions and allowing us to witness the cycle of waste becoming resource firsthand, fostering a deeper connection to our consumption and its impacts.

However, to hold these hopeful projections as the sole solution would be misguided. The key to truly sustainable recycling lies in reducing the need for it in the first place. This means that the future of recycling also lies in a shift towards decoupling economic growth from resource use - creating less waste to begin with, and a more responsible consumer culture.

Thus, what's crucial is not just the evolution of recycling technology, but a co-evolution of socio-economic systems. A future where conscious consumerism is the norm, with products designed for longevity rather than obsolescence, and industries valuing waste as a resource rather than an unwanted by-product - that is a future that goes beyond recycling.

The responsibility for making this future a reality does not fall on one demographic. Governments, corporations, and individuals all have a role to play. We predict stricter waste and recycling policies coupled with increased transparency about products' environmental impact will encourage companies to innovate and individuals to make informed choices.

Acceptance of Extended Producer Responsibility (EPR) practices can underpin this future where producers consider the entire lifecycle of the products they design, produce, and sell. This would mean designing for recyclability, offering take-back programs, and educating the consumer about their product's environmental footprint.

The participation of every individual in waste reduction will also be a cornerstone of this future. Everyone's involvement can accelerate the transition towards waste valorization, where waste is seen as a valuable asset instead of a burden.

In addition, we imagine a future where recycling is more than just managing waste but a tool for social equity too. Programs that incentivize the collection and drop off of recyclable materials can help clear litter in public spaces while simultaneously providing a source of income for the needy.

Furthermore, the educational aspect of recycling should not be overlooked. As recycling evolves, so too should our understanding of its importance. Clear, accessible information about best recycling practices, the impact of improper disposal, and the real cost of our consumption is necessary.

The future of recycling is bright and promising, filled with contours of technological breakthroughs, evolved systems, and enlightened consumer habits. However, to bring this future into existence, each of us must take responsibility for our own role in the waste cycle, constantly learning and adapting, and pledging to prioritize the health of our shared home - planet Earth.

In conclusion, while these predictions and possibilities for recycling are encouraging, they can't be realized without conscious effort and commitment across all levels of society. The key to a future of sustainable recycling is not only in the progression of technology but in our attitudes towards consumption and waste. The choices we make today will shape the state of recycling - and, by extension, the state of the planet - in the years to come.

Closing Remarks:
An Invitation to Act and Transform

As we come to the end of this exploration into the intricacies of climate change and the modern methods of recycling, we're led to the realization of a collective responsibility. The facts of climate change are undeniable, and our wasteful tendencies contribute significantly to it. While the repercussions may be daunting, the good news is that there are tangible actions that we can take to curb this trend and transform our future.

Consider every single piece of plastic you've encountered and the food you've wasted. Reflect on how our systems of production and consumption – once thought unassailable – can be reshaped into a circular economy. Integrating revolutionary methods of recycling, informed by modern science and technology, can bring about significant changes. Technology now affords us actionable solutions that can reduce, reuse, and recycle our waste more efficiently than ever before.

While we must be conscientious about our personal consumption and waste production, our efforts don't stop at our doorstep. Companies, governments, and even our neighborhoods can and should be held accountable in this transformative journey. Our habits as conscious consumers can influence business policies, regulatory bodies can enforce eco-friendly practices, and gatherings of informed citizens can act as engines for systemic change.

Think of our planet as an inheritance, a precious legacy we have to pass on to future generations. By valuing what would otherwise be discarded, upcycling materials, and extracting worth from waste, we reorient our relationship with the resources of our planet. We must understand that restoration and regeneration form the hallmark of our pledge to the natural world—an investment in recuperative agriculture, supporting renewable energy, and taking the initiative in restoring natural habitats.

As we move forward, this book is your invitation to act and transform. Let much-needed change take root in our lives, and let us together build a progressive planet for future generations. The journey of a thousand miles begins with a single step, and ours start with comprehending the role of recycling in mitigating climate change. Let the predictions and possibilities around recycling be our canvas as we paint the future of sustainability. Let's create a better tomorrow – beginning today.

Appendix A:
Resources and Extra Reading

As we draw to a close in our exploration of recycling and climate change, there's still much to learn. This appendix is here to guide you to further expand your understanding and to inspire you to become an active participant in the fight against climate change. It offers suggested resources and extra readings that are well-researched and influential in their respective fields.

Web Resources

The Intergovernmental Panel on Climate Change (IPCC) provides scientific information about the implications of climate change and potential future risks. Their reports provide comprehensive evaluations of current research.

The Environmental Protection Agency (EPA) covers a wide variety of topics around climate change and waste management, offering guides, facts, and data.

Recycling International offers global news on recycling trends, innovative technologies, regulations, and other relevant topics.

Books

A lot of great books are exploring these topics in depth. Here are a few that will expand your knowledge and understanding:

This Changes Everything: Capitalism vs. The Climate by Naomi Klein - An analysis of how economic changes can positively impact the climate.

Waste: Uncovering the Global Food Scandal by Tristram Stuart - A discourse on food waste and its impact on climate change.

Cradle to Cradle: Remaking the Way We Make Things by William McDonough and Michael Braungart - Explores the concept of circular economy and how it can revolutionize the way we view waste.

The Story of Stuff by Annie Leonard - Discusses the lifecycle of material goods, from production, through our homes, and then to disposal.

Documentaries

Visual learning can be a powerful tool. Here are some enlightening documentaries about climate change and recycling:

"An Inconvenient Truth" - Former Vice President Al Gore's campaign to educate the public about the severity of climate change.

"The Story of Plastic" - A comprehensive look at the man-made crisis of plastic pollution and the global movement that's rising to solve it.

"Plastic China" - An insightful documentary that shows the impact of our global plastic consumption on the lives of individuals in China.

While this is not an exhaustive list, these resources will provide you with ample material to delve deeper into the topics we've discussed throughout this book. Lastly, remember that knowledge is power, but it is just the first step. The true power lies in using this knowledge to influence your choices and actions towards a more sustainable future.